Immunology

Immunology

Edited by
Joffrey Butler

⊟ Larsen & Keller
www.larsen-keller.com

Immunology
Edited by Joffrey Butler
ISBN: 978-1-63549-148-7 (Hardback)

⊟ Larsen & Keller

Published by Larsen and Keller Education,
5 Penn Plaza,
19th Floor,
New York, NY 10001, USA

Cataloging-in-Publication Data

Immunology / edited by Joffrey Butler.
    p. cm.
Includes bibliographical references and index.
ISBN 978-1-63549-148-7
1. Immunology. 2. Immune system. 3. Immunodiagnosis. I. Butler, Joffrey.
QR181 .I46 2017
616.079--dc23

This book contains information obtained from authentic and highly regarded sources. All chapters are published with permission under the Creative Commons Attribution Share Alike License or equivalent. A wide variety of references are listed. Permissions and sources are indicated; for detailed attributions, please refer to the permissions page. Reasonable efforts have been made to publish reliable data and information, but the authors, editors and publisher cannot assume any responsibility for the vailidity of all materials or the consequences of their use.

Trademark Notice: All trademarks used herein are the property of their respective owners. The use of any trademark in this text does not vest in the author or publisher any trademark ownership rights in such trademarks, nor does the use of such trademarks imply any affiliation with or endorsement of this book by such owners.

The publisher's policy is to use permanent paper from mills that operate a sustainable forestry policy. Furthermore, the publisher ensures that the text paper and cover boards used have met acceptable environmental accreditation standards.

Printed and bound in the United States of America.

For more information regarding Larsen and Keller Education and its products, please visit the publisher's website www.larsen-keller.com

Table of Contents

Preface

Immunology refers to the study of immune systems of all living organisms. It studies the physiological functions, immunological disorders and the chemical, physical and physiological characteristics of immune systems. This book is a valuable compilation of topics, ranging from the basic to the most complex theories and principles in this field. Different approaches, evaluations and methodologies and advanced studies on immunology have been included in it. The text aims to serve as a resource guide for students and facilitate the study of the discipline.

A foreword of all chapters of the book is provided below:

Chapter 1 - A branch of biomedical science, immunology is the study of immune systems in all organisms. It catalogues, measures and inspects the physiological functioning of the body in states of health and disease. This chapter provides valuable information about the discipline of immunology, immune systems and aspects of immunity- humoral immunity and cell-mediated immunity; **Chapter 2 -** The branches of immunology are based on the application of immunity research to various fields. This chapter informs the reader about branches like cancer immunology, reproductive immunology, testicular immunology, osteoimmunology, psychoneuroimmunology, immunoproteomics etc. The reader is provided with in-depth knowledge of each of these branches in this chapter; **Chapter 3 -** This chapter deals exclusively with the immune system and its categories. The chapter provides detailed information about the innate immune system, adaptive immune system and complement system. The chapter acquaints the reader with the definition and discerning characteristics of each division; **Chapter 4 -** The lymphatic system is a part of the circulatory system and a vital part of the immune system consists of a network of lymphatic vessels that carry a clear fluid called lymph toward the heart. This chapter studies the lymphatic system and the organs and tissues that comprise it. Also in this chapter is a description of the role of lymph in the immune function of the body; **Chapter 5 -** The immune system can fall prey to malfunctioning and disorders of the immune system fall into three broad categories – immunodeficiency, autoimmunity and hypersensitivities. The chapter explores each of these disorders and provides an integrated study of each. A section of the chapter specifically deals with transplant rejection, HIV and AIDS and chronic granulomatous disease; **Chapter 6 -** There are a plethora of options to treat immunological disorders. This chapter discusses the topic of immunodiagnostics that utilizes immunoassays to measure the presence or concentration of a macromolecule or a small molecule in a solution through the use of an antigen or an antibody. Another method of diagnostics is immunotherapy that is either designed to elicit /amplify the immune response or to reduce/ suppress an immune response as the situation calls for. The aspects of each option elucidated in the chapter are of vital importance and promote a better understanding of the topic of the diagnosis and treatment of immunological disorders.

At the end, I would like to thank all the people associated with this book devoting their precious time and providing their valuable contributions to this book. I would also like to express my gratitude to my fellow colleagues who encouraged me throughout the process.

Editor

Introduction to Immunology

A branch of biomedical science, immunology is the study of immune systems in all organisms. It catalogues, measures and inspects the physiological functioning of the body in states of health and disease. This chapter provides valuable information about the discipline of immunology, immune systems and aspects of immunity- humoral immunity and cell-mediated immunity.

Immunology

Immunology is a branch of biomedical science that covers the study of immune systems in all organisms. It charts, measures, and contextualizes the: physiological functioning of the immune system in states of both health and diseases; malfunctions of the immune system in immunological disorders (such as autoimmune diseases, hypersensitivities, immune deficiency, and transplant rejection); the physical, chemical and physiological characteristics of the components of the immune system *in vitro*, *in situ*, and *in vivo*. Immunology has applications in numerous disciplines of medicine, particularly in the fields of organ transplantation, oncology, virology, bacteriology, parasitology, psychiatry, and dermatology.

Prior to the designation of immunity from the etymological root *immunis*, which is Latin for "exempt"; early physicians characterized organs that would later be proven as essential components of the immune system. The important lymphoid organs of the immune system are the thymus and bone marrow, and chief lymphatic tissues such as spleen, tonsils, lymph vessels, lymph nodes, adenoids, and liver. When health conditions worsen to emergency status, portions of immune system organs including the thymus, spleen, bone marrow, lymph nodes and other lymphatic tissues can be surgically excised for examination while patients are still alive.

Many components of the immune system are typically cellular in nature and not associated with any specific organ; but rather are embedded or circulating in various tissues located throughout the body.

Classical Immunology

Classical immunology ties in with the fields of epidemiology and medicine. It studies the relationship between the body systems, pathogens, and immunity. The earliest written mention of immunity can be traced back to the plague of Athens in 430 BCE. Thucydides noted that people who had recovered from a previous bout of the disease could nurse the sick without contracting the illness a second time. Many other ancient societies have references to this phenomenon, but it was not until the 19th and 20th centuries before the concept developed into scientific theory.

The study of the molecular and cellular components that comprise the immune system, including their function and interaction, is the central science of immunology. The immune system has been divided into a more primitive innate immune system and, in vertebrates, an acquired or adaptive immune system. The latter is further divided into humoral (or antibody) and cell-mediated components.

The humoral (antibody) response is defined as the interaction between antibodies and antigens. Antibodies are specific proteins released from a certain class of immune cells known as B lymphocytes, while antigens are defined as anything that elicits the generation of antibodies ("anti"body "gen"erators). Immunology rests on an understanding of the properties of these two biological entities and the cellular response to both.

Immunological research continues to become more specialized, pursuing non-classical models of immunity and functions of cells, organs and systems not previously associated with the immune system (Yemeserach 2010).

Clinical Immunology

Clinical immunology is the study of diseases caused by disorders of the immune system (failure, aberrant action, and malignant growth of the cellular elements of the system). It also involves diseases of other systems, where immune reactions play a part in the pathology and clinical features.

The diseases caused by disorders of the immune system fall into two broad categories:

- immunodeficiency, in which parts of the immune system fail to provide an adequate response (examples include chronic granulomatous disease and primary immune diseases);
- autoimmunity, in which the immune system attacks its own host's body (examples include systemic lupus erythematosus, rheumatoid arthritis, Hashimoto's disease and myasthenia gravis).

Other immune system disorders include various hypersensitivities (such as in asthma and other allergies) that respond inappropriately to otherwise harmless compounds.

The most well-known disease that affects the immune system itself is AIDS, an immunodeficiency characterized by the suppression of CD4+ ("helper") T cells, dendritic cells and macrophages by the Human Immunodeficiency Virus (HIV).

Clinical immunologists also study ways to prevent the immune system's attempts to destroy allografts (transplant rejection).

Developmental Immunology

The body's capability to react to antigen depends on a person's age, antigen type, maternal factors and the area where the antigen is presented. Neonates are said to be in a state of physiological immunodeficiency, because both their innate and adaptive immunological responses are greatly suppressed. Once born, a child's immune system responds favorably to protein antigens while not as well to glycoproteins and polysaccharides. In fact, many of the infections acquired by neonates are caused by low virulence organisms like *Staphylococcus* and *Pseudomonas*. In neonates, opsonic activity and the ability to activate the complement cascade is very limited. For example, the mean level of C3 in a newborn is approximately 65% of that found in the adult. Phagocytic activity is also greatly impaired in newborns. This is due to lower opsonic activity, as well as diminished up-regulation of integrin and selectin receptors, which limit the ability of neutrophils to interact with adhesion molecules in the endothelium. Their monocytes are slow and have a reduced ATP production, which also limits the newborn's phagocytic activity. Although, the number of total lymphocytes is significantly higher than in adults, the cellular and humoral immunity is also impaired. Antigen-presenting cells in newborns have a reduced capability to activate T cells. Also, T cells of a newborn proliferate poorly and produce very small amounts of cytokines like IL-2, IL-4, IL-5, IL-12, and IFN-g which limits their capacity to activate the humoral response as well as the phagocitic activity of macrophage. B cells develop early during gestation but are not fully active.

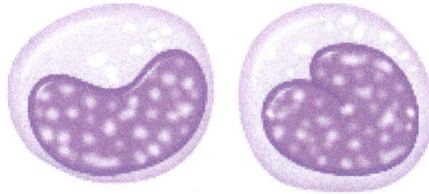

Artist's impression of monocytes.

Maternal factors also play a role in the body's immune response. At birth, most of the immunoglobulin present is maternal IgG. Because IgM, IgD, IgE and IgA don't cross the placenta, they are almost undetectable at birth. Some IgA is provided by breast milk. These passively-acquired antibodies can protect the newborn for up to 18 months, but their response is usually short-lived and of low affinity. These antibodies can also produce a negative response. If a child is exposed to the antibody for a particular antigen before being exposed to the antigen itself then the child will produce a dampened response. Passively acquired maternal antibodies can suppress the antibody response to active immunization. Similarly the response of T-cells to vaccination differs in children compared to adults, and vaccines that induce Th1 responses in adults do not readily elicit these same responses in neonates. Between six and nine months after birth, a child's immune system begins to respond more strongly to glycoproteins, but there is usually no marked improvement in their response to polysaccharides until they are at least one year old. This can be the reason for distinct time frames found in vaccination schedules.

During adolescence, the human body undergoes various physical, physiological and immunological changes triggered and mediated by hormones, of which the most significant in females is 17-β-oestradiol (an oestrogen) and, in males, is testosterone. Oestradiol usually begins to act around the age of 10 and testosterone some months later. There is evidence that these steroids act directly not only on the primary and secondary sexual characteristics but also have an effect on the development and regulation of the immune system, including an increased risk in developing pubescent and post-pubescent autoimmunity. There is also some evidence that cell surface receptors on B cells and macrophages may detect sex hormones in the system.

The female sex hormone 17-β-oestradiol has been shown to regulate the level of immunological response, while some male androgens such as testosterone seem to suppress the stress response to infection. Other androgens, however, such as DHEA, increase immune response. As in females, the male sex hormones seem to have more control of the immune system during puberty and post-puberty than during the rest of a male's adult life.

Physical changes during puberty such as thymic involution also affect immunological response.

Immunotherapy

The use of immune system components to treat a disease or disorder is known as immunotherapy. Immunotherapy is most commonly used in the context of the treatment of cancers together with chemotherapy (drugs) and radiotherapy (radiation). However, immunotherapy is also often used in the immunosuppressed (such as HIV patients) and people suffering from other immune deficiencies or autoimmune diseases. This includes regulating factors such as IL-2, IL-10, GM-CSF B, IFN-α.

Diagnostic Immunology

The specificity of the bond between antibody and antigen has made the antibody an excellent tool for the detection of substances by a variety of diagnostic techniques. Antibodies specific for a desired antigen can be conjugated with an isotopic (radio) or fluorescent label or with a color-forming enzyme in order to detect it. However, the similarity between some antigens can lead to false positives and other errors in such tests by antibodies cross-reacting with antigens that aren't exact matches.

Cancer Immunology

The study of the interaction of the immune system with cancer cells can lead to diagnostic tests and therapies with which to find and fight cancer.

Reproductive Immunology

This area of the immunology is devoted to the study of immunological aspects of the reproductive process including fetus acceptance. The term has also been used by fertility clinics to address fertility problems, recurrent miscarriages, premature deliveries and dangerous complications such as pre-eclampsia.

Theoretical Immunology

Immunology is strongly experimental in everyday practice but is also characterized by an ongoing theoretical attitude. Many theories have been suggested in immunology from the end of the nineteenth century up to the present time. The end of the 19th century and the beginning of the 20th century saw a battle between "cellular" and "humoral" theories of immunity. According to the cellular theory of immunity, represented in particular by Elie Metchnikoff, it was cells – more precisely, phagocytes – that were responsible for immune responses. In contrast, the humoral theory of immunity, held by Robert Koch and Emil von Behring, among others, stated that the active immune agents were soluble components (molecules) found in the organism's "humors" rather than its cells.

In the mid-1950s, Frank Burnet, inspired by a suggestion made by Niels Jerne, formulated the clonal selection theory (CST) of immunity. On the basis of CST, Burnet developed a theory of how an immune response is triggered according to the self/nonself distinction: "self" constituents (constituents of the body) do not trigger destructive immune responses, while "nonself" entities (e.g., pathogens, an allograft) trigger a destructive immune response. The theory was later modified to reflect new discoveries regarding histocompatibility or the complex "two-signal" activation of T cells. The self/nonself theory of immunity and the self/nonself vocabulary have been criticized, but remain very influential.

More recently, several theoretical frameworks have been suggested in immunology, including "autopoietic" views, "cognitive immune" views, the "danger model" (or "danger theory"), and the "discontinuity" theory. The danger model, suggested by Polly Matzinger and colleagues, has been very influential, arousing many comments and discussions.

Immunologist

According to the American Academy of Allergy, Asthma, and Immunology (AAAAI), "an immunologist is a research scientist who investigates the immune system of vertebrates (including the human immune system). Immunologists include research scientists (PhDs) who work in laboratories. Immunologists also include physicians who, for example, treat patients with immune system disorders. Some immunologists are physician-scientists who combine laboratory research with patient care."

Career in Immunology

Bioscience is the overall major in which undergraduate students who are interested in general well-being take in college. Immunology is a branch of bioscience for undergraduate programs but the major gets specified as students move on for graduate program in immunology. The aim of immunology is to study the health of humans and animals through effective yet consistent research, (AAAAI, 2013). The most important thing about being immunologists is the research because it is the biggest portion of their jobs.

Most graduate immunology schools follow the AAI courses immunology which are offered throughout numerous schools in the United States. For example, in New York State, there are several universities that offer the AAI courses immunology: Albany Medical College, Cornell University, Icahn School of Medicine at Mount Sinai, New York University Langone Medical Center, University at Albany (SUNY), University at Buffalo (SUNY), University of Rochester Medical Center and Upstate Medical University (SUNY). The AAI immunology courses include an Introductory Course and an Advance Course. The Introductory Course is a course that gives students an overview of the basics of immunology.

In addition, this Introductory Course gives students more information to complement general biology or science training. It also has two different parts: Part I is an introduction to the basic principles of immunology and Part II is a clinically-oriented lecture series. On the other hand, the Advanced Course is another course for those who are willing to expand or update their understanding of immunology. It is advised for students who want to attend the Advanced Course to have a background of the principles of immunology. Most schools require students to take electives in other to complete their degrees. A Master's degree requires two years of study following the attainment of a bachelor's degree. For a doctoral programme it is required to take two additional years of study.

The expectation of occupational growth in immunology is an increase of 36 percent from 2010 to 2020. The median annual wage was $76,700 in May 2010. However, the lowest 10 percent of immunologists earned less than $41,560, and the top 10 percent earned more than $142,800, (Bureau of Labor Statistics, 2013). The practice of immunology itself is not specified by the U.S. Department of Labor but it belongs to the practice of life science in general.

Immune System

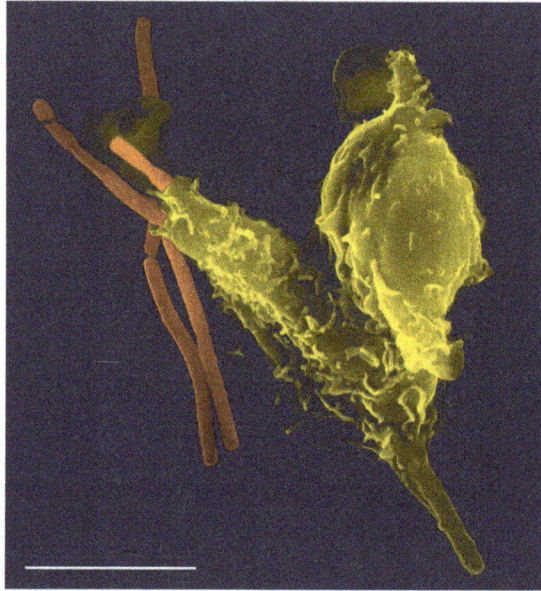

A scanning electron microscope image of a single neutrophil (yellow), engulfing anthrax bacteria (orange)

The immune system is a host defense system comprising many biological structures and processes within an organism that protects against disease. To function properly, an immune system must detect a wide variety of agents, known as pathogens, from viruses to parasitic worms, and distinguish them from the organism's own healthy tissue. In many species, the immune system can be classified into subsystems, such as the innate immune system versus the adaptive immune system, or humoral immunity versus cell-mediated immunity. In humans, the blood–brain barrier, blood–cerebrospinal fluid barrier, and similar fluid–brain barriers separate the peripheral immune system from the neuroimmune system which protects the brain.

Pathogens can rapidly evolve and adapt, and thereby avoid detection and neutralization by the immune system; however, multiple defense mechanisms have also evolved to recognize and neutralize pathogens. Even simple unicellular organisms such as bacteria possess a rudimentary immune system, in the form of enzymes that protect against bacteriophage infections. Other basic immune mechanisms evolved in ancient eukaryotes and remain in their modern descendants, such as plants and invertebrates. These mechanisms include phagocytosis, antimicrobial peptides called defensins, and the complement system. Jawed vertebrates, including humans, have even more sophisticated defense mechanisms, including the ability to adapt over time to recognize specific pathogens more efficiently. Adaptive (or acquired) immunity creates immunological memory after an initial response to a specific pathogen, leading to an enhanced response to subsequent encounters with that same pathogen. This process of acquired immunity is the basis of vaccination.

Disorders of the immune system can result in autoimmune diseases, inflammatory diseases and cancer. Immunodeficiency occurs when the immune system is less active than normal, resulting in recurring and life-threatening infections. In humans, immunodeficiency can either be the result of a genetic disease such as severe combined immunodeficiency, acquired conditions such as HIV/AIDS, or the use of immunosuppressive medication. In contrast, autoimmunity results from a hyperactive immune system attacking normal tissues as if they were foreign organisms. Common autoimmune diseases include Hashimoto's thyroiditis, rheumatoid arthritis, diabetes mellitus type 1, and systemic lupus erythematosus. Immunology covers the study of all aspects of the immune system.

History of Immunology

Immunology is a science that examines the structure and function of the immune system. It originates from medicine and early studies on the causes of immunity to disease. The earliest known reference to immunity was during the plague of Athens in 430 BC. Thucydides noted that people who had recovered from a previous bout of the disease could nurse the sick without contracting the illness a second time. In the 18th century, Pierre-Louis Moreau de Maupertuis made experiments with scorpion venom and observed that certain dogs and mice were immune to this venom. This and other observations of acquired immunity were later exploited by Louis Pasteur in his development of vaccination and his proposed germ theory of disease. Pasteur's theory was in direct opposition to contemporary theories of disease, such as the miasma theory. It was not until Robert Koch's 1891 proofs, for which he was awarded a Nobel Prize in 1905, that microorganisms were confirmed as the cause of infectious disease. Viruses were confirmed as human pathogens in 1901, with the discovery of the yellow fever virus by Walter Reed.

Immunology made a great advance towards the end of the 19th century, through rapid developments, in the study of humoral immunity and cellular immunity. Particularly important was the work of Paul Ehrlich, who proposed the side-chain theory to explain the specificity of the antigen-antibody reaction; his contributions to the understanding of humoral immunity were recognized by the award of a Nobel Prize in 1908, which was jointly awarded to the founder of cellular immunology, Elie Metchnikoff.

Layered Defense

The immune system protects organisms from infection with layered defenses of increasing specificity. In simple terms, physical barriers prevent pathogens such as bacteria and viruses from entering the organism. If a pathogen breaches these barriers, the innate immune system provides an immediate, but non-specific response. Innate immune systems are found in all plants and animals. If pathogens successfully evade the innate response, vertebrates possess a second layer of protection, the adaptive immune system, which is activated by the innate response. Here, the immune system adapts its response during an infection to improve its recognition of the pathogen. This improved response is then retained after the pathogen has been eliminated, in the form of an immunological memory, and allows the adaptive immune system to mount faster and stronger attacks each time this pathogen is encountered.

Components of the immune system	
Innate immune system	Adaptive immune system
Response is non-specific	Pathogen and antigen specific response
Exposure leads to immediate maximal response	Lag time between exposure and maximal response
Cell-mediated and humoral components	Cell-mediated and humoral components
No immunological memory	Exposure leads to immunological memory
Found in nearly all forms of life	Found only in jawed vertebrates

Both innate and adaptive immunity depend on the ability of the immune system to distinguish between self and non-self molecules. In immunology, self molecules are those components of an organism's body that can be distinguished from foreign substances by the immune system. Conversely, non-self molecules are those recognized as foreign molecules. One class of non-self molecules are called antigens (short for antibody generators) and are defined as substances that bind to specific immune receptors and elicit an immune response.

Innate Immune System

Microorganisms or toxins that successfully enter an organism encounter the cells and mechanisms of the innate immune system. The innate response is usually triggered when microbes are identified by pattern recognition receptors, which recognize components that are conserved among broad groups of microorganisms, or when damaged, injured or stressed cells send out alarm signals, many of which (but not all) are recognized by the same receptors as those that recognize pathogens. Innate immune defenses are non-specific, meaning these systems respond to pathogens in a generic way. This system does not confer long-lasting immunity against a pathogen. The innate immune system is the dominant system of host defense in most organisms.

Surface Barriers

Several barriers protect organisms from infection, including mechanical, chemical, and biological barriers. The waxy cuticle of many leaves, the exoskeleton of insects, the shells and membranes of externally deposited eggs, and skin are examples of mechanical barriers that are the first line of defense against infection. However, as organisms cannot be completely sealed from their environments, other systems act to protect body openings such as the lungs, intestines, and the genitourinary tract. In the lungs, coughing and sneezing mechanically eject pathogens and other irritants from the respiratory tract. The flushing action of tears and urine also mechanically expels pathogens, while mucus secreted by the respiratory and gastrointestinal tract serves to trap and entangle microorganisms.

Chemical barriers also protect against infection. The skin and respiratory tract secrete antimicrobial peptides such as the β-defensins. Enzymes such as lysozyme and phospholipase A2 in saliva, tears, and breast milk are also antibacterials. Vaginal secretions serve as a chemical barrier following menarche, when they become slightly acidic, while semen contains defensins and zinc to kill pathogens. In the stomach, gastric acid and proteases serve as powerful chemical defenses against ingested pathogens.

Within the genitourinary and gastrointestinal tracts, commensal flora serve as biological barriers by competing with pathogenic bacteria for food and space and, in some cases, by changing the conditions in their environment, such as pH or available iron. This reduces the probability that pathogens will reach sufficient numbers to cause illness. However, since most antibiotics non-specifically target bacteria and do not affect fungi, oral antibiotics can lead to an "overgrowth" of fungi and cause conditions such as a vaginal candidiasis (a yeast infection). There is good evidence that re-introduction of probiotic flora, such as pure cultures of the lactobacilli normally found in unpasteurized yogurt, helps restore a healthy balance of microbial populations in intestinal infections in children and encouraging preliminary data in studies on bacterial gastroenteritis, inflammatory bowel diseases, urinary tract infection and post-surgical infections.

Inflammation

Inflammation is one of the first responses of the immune system to infection. The symptoms of inflammation are redness, swelling, heat, and pain, which are caused by increased blood flow into tissue. Inflammation is produced by eicosanoids and cytokines, which are released by injured or infected cells. Eicosanoids include prostaglandins that produce fever and the dilation of blood vessels associated with inflammation, and leukotrienes that attract certain white blood cells (leukocytes). Common cytokines include interleukins that are responsible for communication between white blood cells; chemokines that promote chemotaxis; and interferons that have anti-viral effects, such as shutting down protein synthesis in the host cell. Growth factors and cytotoxic factors may also be released. These cytokines and other chemicals recruit immune cells to the site of infection and promote healing of any damaged tissue following the removal of pathogens.

Complement System

The complement system is a biochemical cascade that attacks the surfaces of foreign cells. It contains over 20 different proteins and is named for its ability to "complement" the killing of pathogens by antibodies. Complement is the major humoral component of the innate immune response. Many species have complement systems, including non-mammals like plants, fish, and some invertebrates.

In humans, this response is activated by complement binding to antibodies that have attached to these microbes or the binding of complement proteins to carbohydrates on the surfaces of microbes. This recognition signal triggers a rapid killing response. The speed of the response is a result of signal amplification that occurs following sequential proteolytic activation of complement molecules, which are also proteases. After complement proteins initially bind to the microbe, they activate their protease activity, which in turn activates other complement proteases, and so on. This produces a catalytic cascade that amplifies the initial signal by controlled positive feedback. The cascade results in the production of peptides that attract immune cells, increase vascular permeability, and opsonize (coat) the surface of a pathogen, marking it for destruction. This deposition of complement can also kill cells directly by disrupting their plasma membrane.

Cellular Barriers

Leukocytes (white blood cells) act like independent, single-celled organisms and are the second arm of the innate immune system. The innate leukocytes include the phagocytes (macrophages, neutrophils, and dendritic cells), innate lymphoid cells, mast cells, eosinophils, basophils, and natural killer cells. These cells identify and eliminate pathogens, either by attacking larger pathogens through contact or by engulfing and then killing microorganisms. Innate cells are also important mediators in lymphoid organ development and the activation of the adaptive immune system.

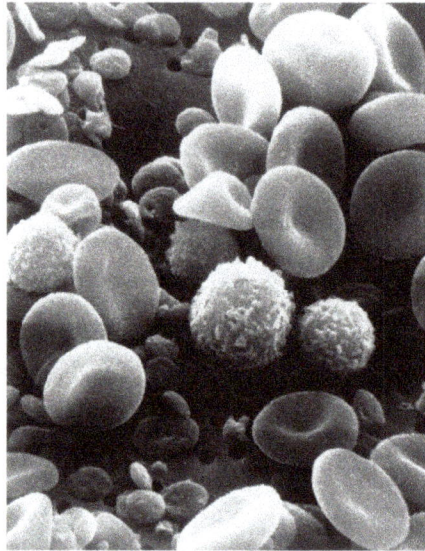

A scanning electron microscope image of normal circulating human blood. One can see red blood cells, several knobby white blood cells including lymphocytes, a monocyte, a neutrophil, and many small disc-shaped platelets.

Phagocytosis is an important feature of cellular innate immunity performed by cells called 'phagocytes' that engulf, or eat, pathogens or particles. Phagocytes generally patrol the body searching for pathogens, but can be called to specific locations by cytokines. Once a pathogen has been engulfed by a phagocyte, it becomes trapped in an intracellular vesicle called a phagosome, which subsequently fuses with another vesicle called a lysosome to form a phagolysosome. The pathogen is killed by the activity of digestive enzymes or following a respiratory burst that releases free radicals into the phagolysosome. Phagocytosis evolved as a means of acquiring nutrients, but this role was extended in phagocytes to include engulfment of pathogens as a defense mechanism. Phagocytosis probably represents the oldest form of host defense, as phagocytes have been identified in both vertebrate and invertebrate animals.

Neutrophils and macrophages are phagocytes that travel throughout the body in pursuit of invading pathogens. Neutrophils are normally found in the bloodstream and are the most abundant type of phagocyte, normally representing 50% to 60% of the total circulating leukocytes. During the acute phase of inflammation, particularly as a result of bacterial infection, neutrophils migrate toward the site of inflammation in a process called chemotaxis, and are usually the first cells to arrive at the scene of infection. Macrophages are versatile cells that reside within tissues and: (i) produce a wide array of chemicals including enzymes, complement proteins, and cytokines, while they can also (ii) act as scavengers that rid the body of worn-out cells and other debris, and as antigen-presenting cells that activate the adaptive immune system.

Dendritic cells (DC) are phagocytes in tissues that are in contact with the external environment; therefore, they are located mainly in the skin, nose, lungs, stomach, and intestines. They are named for their resemblance to neuronal dendrites, as both have many spine-like projections, but dendritic cells are in no way connected to the nervous system. Dendritic cells serve as a link between the bodily tissues and the innate and adaptive immune systems, as they present antigens to T cells, one of the key cell types of the adaptive immune system.

Mast cells reside in connective tissues and mucous membranes, and regulate the inflammatory response. They are most often associated with allergy and anaphylaxis. Basophils and eosinophils are related to neutrophils. They secrete chemical mediators that are involved in defending against parasites and play a role in allergic reactions, such as asthma. Natural killer (NK cells) cells are leukocytes that attack and destroy tumor cells, or cells that have been infected by viruses.

Natural Killer Cells

Natural killer cells, or NK cells, are a component of the innate immune system which does not directly attack invading microbes. Rather, NK cells destroy compromised host cells, such as tumor cells or virus-infected cells, recognizing such cells by a condition known as "missing self." This term describes cells with low levels of a cell-surface marker called MHC I (major histocompatibility complex) – a situation that can arise in viral infections of host cells. They were named "natural killer" because of the initial notion that they do not require activation in order to kill cells that are "missing self." For many years it was unclear how NK cells recognize tumor cells and infected cells. It is now known that the MHC makeup on the surface of those cells is altered and the NK cells become activated through recognition of "missing self". Normal body cells are not recognized and attacked by NK cells because they express intact self MHC antigens. Those MHC antigens are recognized by killer cell immunoglobulin receptors (KIR) which essentially put the brakes on NK cells.

Adaptive Immune System

The adaptive immune system evolved in early vertebrates and allows for a stronger immune response as well as immunological memory, where each pathogen is "remembered" by a signature antigen. The adaptive immune response is antigen-specific and requires the recognition of specific "non-self" antigens during a process called antigen presentation. Antigen specificity allows for the generation of responses that are tailored to specific pathogens or pathogen-infected cells. The ability to mount these tailored responses is maintained in the body by "memory cells". Should a pathogen infect the body more than once, these specific memory cells are used to quickly eliminate it.

Lymphocytes

The cells of the adaptive immune system are special types of leukocytes, called lymphocytes. B cells and T cells are the major types of lymphocytes and are derived from hematopoietic stem cells in the bone marrow. B cells are involved in the humoral immune response, whereas T cells are involved in cell-mediated immune response.

Both B cells and T cells carry receptor molecules that recognize specific targets. T cells recognize a "non-self" target, such as a pathogen, only after antigens (small fragments of the pathogen) have been processed and presented in combination with a "self" receptor called a major histocompatibility complex (MHC) molecule. There are two major subtypes of T cells: the killer T cell and the helper T cell. In addition there are regulatory T cells which have a role in modulating immune response. Killer T cells only recognize antigens coupled to Class I MHC molecules, while helper T cells and regulatory T cells only recognize antigens coupled to Class II MHC molecules. These two mechanisms of antigen presentation reflect the different roles of the two types of T cell. A third, minor subtype are the γδ T cells that recognize intact antigens that are not bound to MHC receptors.

In contrast, the B cell antigen-specific receptor is an antibody molecule on the B cell surface, and recognizes whole pathogens without any need for antigen processing. Each lineage of B cell expresses a different antibody, so the complete set of B cell antigen receptors represent all the antibodies that the body can manufacture.

Killer T Cells

Killer T cells are a sub-group of T cells that kill cells that are infected with viruses (and other pathogens), or are otherwise damaged or dysfunctional. As with B cells, each type of T cell recognizes a different antigen. Killer T cells are activated when their T cell receptor (TCR) binds to this specific antigen in a complex with the MHC Class I receptor of another cell. Recognition of this MHC:antigen complex is aided by a co-receptor on the T cell, called CD8. The T cell then travels throughout the body in search of cells where the MHC I receptors bear this antigen. When an activated T cell contacts such cells, it releases cytotoxins, such as perforin, which form pores in the target cell's plasma membrane, allowing ions, water and toxins to enter. The entry of another toxin called granulysin (a protease) induces the target cell to undergo apoptosis. T cell killing of host cells is particularly important in preventing the replication of viruses. T cell activation is tightly controlled and generally requires a very strong MHC/antigen activation signal, or additional activation signals provided by "helper" T cells.

Helper T Cells

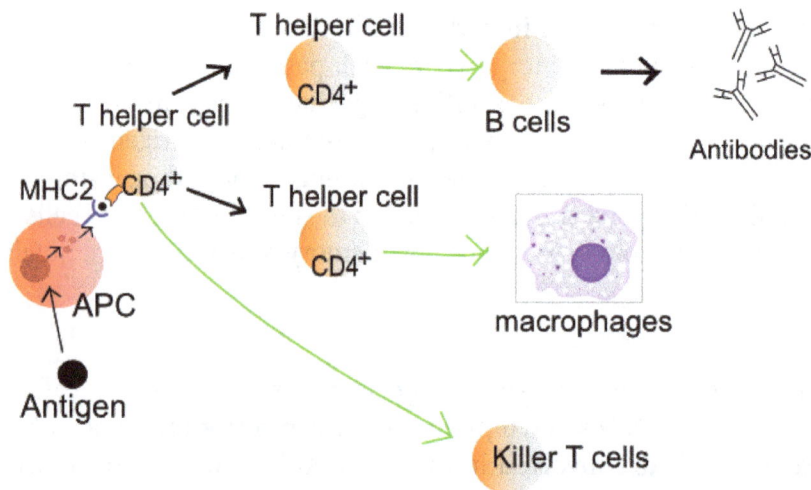

Function of T helper cells: Antigen-presenting cells (APCs) present antigen on their Class II MHC molecules (MHC2). Helper T cells recognize these, with the help of their expression of CD4 co-receptor (CD4+). The activation of a resting helper T cell causes it to release cytokines and other stimulatory signals (green arrows) that stimulate the activity of macrophages, killer T cells and B cells, the latter producing antibodies. The stimulation of B cells and macrophages succeeds a proliferation of T helper cells.

Helper T cells regulate both the innate and adaptive immune responses and help determine which immune responses the body makes to a particular pathogen. These cells have no cytotoxic activity and do not kill infected cells or clear pathogens directly. They instead control the immune response by directing other cells to perform these tasks.

Helper T cells express T cell receptors (TCR) that recognize antigen bound to Class II MHC molecules. The MHC:antigen complex is also recognized by the helper cell's CD4 co-receptor, which recruits molecules inside the T cell (e.g., Lck) that are responsible for the T cell's activation. Helper T cells have a weaker association with the MHC:antigen complex than observed for killer T cells, meaning many receptors (around 200–300) on the helper T cell must be bound by an MHC:antigen in order to activate the helper cell, while killer T cells can be activated by engagement of a single MHC:antigen molecule. Helper T cell activation also requires longer duration of engagement with an antigen-presenting cell. The activation of a resting helper T cell causes it to release cytokines that influence the activity of many cell types. Cytokine signals produced by helper T cells enhance the microbicidal function of macrophages and the activity of killer T cells. In addition, helper T cell activation causes an upregulation of molecules expressed on the T cell's surface, such as CD40 ligand (also called CD154), which provide extra stimulatory signals typically required to activate antibody-producing B cells.

Gamma Delta T Cells

Gamma delta T cells (γδ T cells) possess an alternative T cell receptor (TCR) as opposed to CD4+ and CD8+ (αβ) T cells and share the characteristics of helper T cells, cytotoxic T cells and NK cells. The conditions that produce responses from γδ T cells are not fully understood. Like other 'unconventional' T cell subsets bearing invariant TCRs, such as CD1d-restricted Natural Killer T cells, γδ T cells straddle the border between innate and adaptive immunity. On one hand, γδ T cells are a component of adaptive immunity as they rearrange TCR genes to produce receptor diversity and can also develop a memory phenotype. On the other hand, the various subsets are also part of the innate immune system, as restricted TCR or NK receptors may be used as pattern recognition receptors. For example, large numbers of human Vγ9/Vδ2 T cells respond within hours to common molecules produced by microbes, and highly restricted Vδ1+ T cells in epithelia respond to stressed epithelial cells.

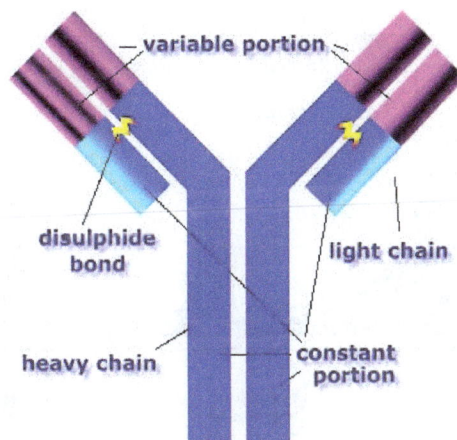

An antibody is made up of two heavy chains and two light chains. The unique variable region allows an antibody to recognize its matching antigen.

B Lymphocytes and Antibodies

A B cell identifies pathogens when antibodies on its surface bind to a specific foreign antigen. This antigen/antibody complex is taken up by the B cell and processed by proteolysis into peptides. The B cell then displays these antigenic peptides on its surface MHC class II molecules. This combination of MHC and antigen attracts a matching helper T cell, which releases lymphokines and activates the B cell. As the activated B cell then begins to divide, its offspring (plasma cells) secrete millions of copies of the antibody that recognizes this antigen. These antibodies circulate in blood plasma and lymph, bind to pathogens expressing the antigen and mark them for destruction by complement activation or for uptake and destruction by phagocytes. Antibodies can also neutralize challenges directly, by binding to bacterial toxins or by interfering with the receptors that viruses and bacteria use to infect cells.

Alternative Adaptive Immune System

Evolution of the adaptive immune system occurred in an ancestor of the jawed vertebrates. Many of the classical molecules of the adaptive immune system (e.g., immunoglobulins and T cell receptors) exist only in jawed vertebrates. However, a distinct lymphocyte-derived molecule has been discovered in primitive jawless vertebrates, such as the lamprey and hagfish. These animals possess a large array of molecules called Variable lymphocyte receptors (VLRs) that, like the antigen receptors of jawed vertebrates, are produced from only a small number (one or two) of genes. These molecules are believed to bind pathogenic antigens in a similar way to antibodies, and with the same degree of specificity.

Immunological Memory

When B cells and T cells are activated and begin to replicate, some of their offspring become long-lived memory cells. Throughout the lifetime of an animal, these memory cells remember each specific pathogen encountered and can mount a strong response if the pathogen is detected again. This is "adaptive" because it occurs during the lifetime of an individual as an adaptation to infection with that pathogen and prepares the immune system for future challenges. Immunological memory can be in the form of either passive short-term memory or active long-term memory.

Passive Memory

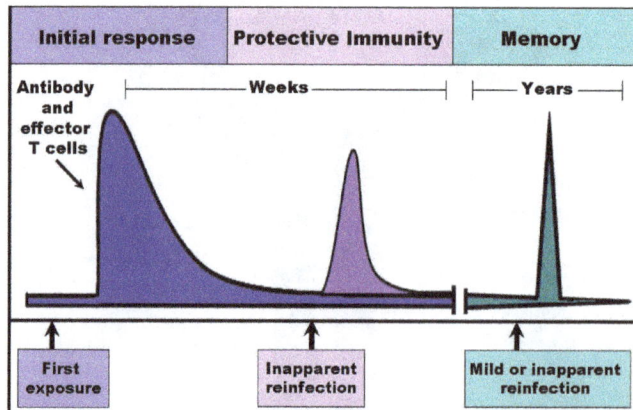

The time-course of an immune response begins with the initial pathogen encounter, (or initial vaccination) and leads to the formation and maintenance of active immunological memory.

Newborn infants have no prior exposure to microbes and are particularly vulnerable to infection. Several layers of passive protection are provided by the mother. During pregnancy, a particular type of antibody, called IgG, is transported from mother to baby directly across the placenta, so human babies have high levels of antibodies even at birth, with the same range of antigen specificities as their mother. Breast milk or colostrum also contains antibodies that are transferred to the gut of the infant and protect against bacterial infections until the newborn can synthesize its own antibodies. This is passive immunity because the fetus does not actually make any memory cells or antibodies—it only borrows them. This passive immunity is usually short-term, lasting from a few days up to several months. In medicine, protective passive immunity can also be transferred artificially from one individual to another via antibody-rich serum.

Active Memory and Immunization

Long-term *active* memory is acquired following infection by activation of B and T cells. Active immunity can also be generated artificially, through vaccination. The principle behind vaccination (also called immunization) is to introduce an antigen from a pathogen in order to stimulate the immune system and develop specific immunity against that particular pathogen without causing disease associated with that organism. This deliberate induction of an immune response is successful because it exploits the natural specificity of the immune system, as well as its inducibility. With infectious disease remaining one of the leading causes of death in the human population, vaccination represents the most effective manipulation of the immune system mankind has developed.

Most viral vaccines are based on live attenuated viruses, while many bacterial vaccines are based on acellular components of micro-organisms, including harmless toxin components. Since many antigens derived from acellular vaccines do not strongly induce the adaptive response, most bacterial vaccines are provided with additional adjuvants that activate the antigen-presenting cells of the innate immune system and maximize immunogenicity.

Disorders of Human Immunity

The immune system is a remarkably effective structure that incorporates specificity, inducibility and adaptation. Failures of host defense do occur, however, and fall into three broad categories: immunodeficiencies, autoimmunity, and hypersensitivities.

Immunodeficiencies

Immunodeficiencies occur when one or more of the components of the immune system are inactive. The ability of the immune system to respond to pathogens is diminished in both the young and the elderly, with immune responses beginning to decline at around 50 years of age due to immunosenescence. In developed countries, obesity, alcoholism, and drug use are common causes of poor immune function. However, malnutrition is the most common cause of immunodeficiency in developing countries. Diets lacking sufficient protein are associated with impaired cell-mediated immunity, complement activity, phagocyte function, IgA antibody concentrations, and cytokine production. Additionally, the loss of the thymus at an early age through genetic mutation or surgical removal results in severe immunodeficiency and a high susceptibility to infection.

Immunodeficiencies can also be inherited or 'acquired'. Chronic granulomatous disease, where phagocytes have a reduced ability to destroy pathogens, is an example of an inherited, or congenital, immunodeficiency. AIDS and some types of cancer cause acquired immunodeficiency.

Autoimmunity

Overactive immune responses comprise the other end of immune dysfunction, particularly the autoimmune disorders. Here, the immune system fails to properly distinguish between self and non-self, and attacks part of the body. Under normal circumstances, many T cells and antibodies react with "self" peptides. One of the functions of specialized cells (located in the thymus and bone marrow) is to present young lymphocytes with self antigens produced throughout the body and to eliminate those cells that recognize self-antigens, preventing autoimmunity.

Hypersensitivity

Hypersensitivity is an immune response that damages the body's own tissues. They are divided into four classes (Type I – IV) based on the mechanisms involved and the time course of the hypersensitive reaction. Type I hypersensitivity is an immediate or anaphylactic reaction, often associated with allergy. Symptoms can range from mild discomfort to death. Type I hypersensitivity is mediated by IgE, which triggers degranulation of mast cells and basophils when cross-linked by antigen. Type II hypersensitivity occurs when antibodies bind to antigens on the patient's own cells, marking them for destruction. This is also called antibody-dependent (or cytotoxic) hypersensitivity, and is mediated by IgG and IgM antibodies. Immune complexes (aggregations of antigens, complement proteins, and IgG and IgM antibodies) deposited in various tissues trigger Type III hypersensitivity reactions. Type IV hypersensitivity (also known as cell-mediated or *delayed type hypersensitivity*) usually takes between two and three days to develop. Type IV reactions are involved in many autoimmune and infectious diseases, but may also involve *contact dermatitis* (poison ivy). These reactions are mediated by T cells, monocytes, and macrophages.

Other Mechanisms and Evolution

It is likely that a multicomponent, adaptive immune system arose with the first vertebrates, as invertebrates do not generate lymphocytes or an antibody-based humoral response. Many species, however, utilize mechanisms that appear to be precursors of these aspects of vertebrate immunity. Immune systems appear even in the structurally most simple forms of life, with bacteria using a unique defense mechanism, called the restriction modification system to protect themselves from viral pathogens, called bacteriophages. Prokaryotes also possess acquired immunity, through a system that uses CRISPR sequences to retain fragments of the genomes of phage that they have come into contact with in the past, which allows them to block virus replication through a form of RNA interference. Offensive elements of the immune systems are also present in unicellular eukaryotes, but studies of their roles in defense are few.

Pattern recognition receptors are proteins used by nearly all organisms to identify molecules associated with pathogens. Antimicrobial peptides called defensins are an evolutionarily conserved component of the innate immune response found in all animals and plants, and represent the main form of invertebrate systemic immunity. The complement system and phagocytic cells are also used by most forms of invertebrate life. Ribonucleases and the RNA interference pathway are conserved across all eukaryotes, and are thought to play a role in the immune response to viruses.

Unlike animals, plants lack phagocytic cells, but many plant immune responses involve systemic chemical signals that are sent through a plant. Individual plant cells respond to molecules associated with pathogens known as Pathogen-associated molecular patterns or PAMPs. When a part of a plant becomes infected, the plant produces a localized hypersensitive response, whereby cells at the site of infection undergo rapid apoptosis to prevent the spread of the disease to other parts of the plant. Systemic acquired resistance (SAR) is a type of defensive response used by plants that renders the entire plant resistant to a particular infectious agent. RNA silencing mechanisms are particularly important in this systemic response as they can block virus replication.

Tumor Immunology

Another important role of the immune system is to identify and eliminate tumors. This is called immune surveillance. The *transformed cells* of tumors express antigens that are not found on normal cells. To the immune system, these antigens appear foreign, and their presence causes immune cells to attack the transformed tumor cells. The antigens expressed by tumors have several sources; some are derived from oncogenic viruses like human papillomavirus, which causes cervical cancer, while others are the organism's own proteins that occur at low levels in normal cells but reach high levels in tumor cells. One example is an enzyme called tyrosinase that, when expressed at high levels, transforms certain skin cells (e.g. melanocytes) into tumors called melanomas. A third possible source of tumor antigens are proteins normally important for regulating cell growth and survival, that commonly mutate into cancer inducing molecules called oncogenes.

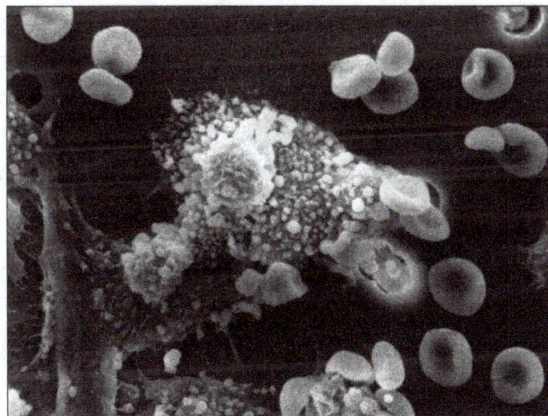

Macrophages have identified a cancer cell (the large, spiky mass). Upon fusing with the cancer cell, the macrophages (smaller white cells) inject toxins that kill the tumor cell. Immunotherapy for the treatment of cancer is an active area of medical research.

The main response of the immune system to tumors is to destroy the abnormal cells using killer T cells, sometimes with the assistance of helper T cells. Tumor antigens are presented on MHC class I molecules in a similar way to viral antigens. This allows killer T cells to recognize the tumor cell as abnormal. NK cells also kill tumorous cells in a similar way, especially if the tumor cells have fewer MHC class I molecules on their surface than normal; this is a common phenomenon with tumors. Sometimes antibodies are generated against tumor cells allowing for their destruction by the complement system.

Clearly, some tumors evade the immune system and go on to become cancers. Tumor cells often have a reduced number of MHC class I molecules on their surface, thus avoiding detection by killer T cells. Some tumor cells also release products that inhibit the immune response; for example by secreting the cytokine TGF-β, which suppresses the activity of macrophages and lymphocytes. In addition, immunological tolerance may develop against tumor antigens, so the immune system no longer attacks the tumor cells.

Paradoxically, macrophages can promote tumor growth when tumor cells send out cytokines that attract macrophages, which then generate cytokines and growth factors that nurture tumor development. In addition, a combination of hypoxia in the tumor and a cytokine produced by macrophages induces tumor cells to decrease production of a protein that blocks metastasis and thereby assists spread of cancer cells.

Physiological Regulation

Hormones can act as immunomodulators, altering the sensitivity of the immune system. For example, female sex hormones are known immunostimulators of both adaptive and innate immune responses. Some autoimmune diseases such as lupus erythematosus strike women preferentially, and their onset often coincides with puberty. By contrast, male sex hormones such as testosterone seem to be immunosuppressive. Other hormones appear to regulate the immune system as well, most notably prolactin, growth hormone and vitamin D.

When a T-cell encounters a foreign pathogen, it extends a vitamin D receptor. This is essentially a signaling device that allows the T-cell to bind to the active form of vitamin D, the steroid hormone calcitriol. T-cells have a symbiotic relationship with vitamin D. Not only does the T-cell extend a vitamin D receptor, in essence asking to bind to the steroid hormone version of vitamin D, calcitriol, but the T-cell expresses the gene CYP27B1, which is the gene responsible for converting the pre-hormone version of vitamin D, calcidiol into the steroid hormone version, calcitriol. Only after binding to calcitriol can T-cells perform their intended function. Other immune system cells that are known to express CYP27B1 and thus activate vitamin D calcidiol, are dendritic cells, keratinocytes and macrophages.

It is conjectured that a progressive decline in hormone levels with age is partially responsible for weakened immune responses in aging individuals. Conversely, some hormones are regulated by the immune system, notably thyroid hormone activity. The age-related decline in immune function is also related to decreasing vitamin D levels in the elderly. As people age, two things happen that negatively affect their vitamin D levels. First, they stay indoors more due to decreased activity levels. This means that they get less sun and therefore produce less cholecalciferol via UVB radiation. Second, as a person ages the skin becomes less adept at producing vitamin D.

Sleep and Rest

The immune system is affected by sleep and rest, and sleep deprivation is detrimental to immune function. Complex feedback loops involving cytokines, such as interleukin-1 and tumor necrosis factor-α produced in response to infection, appear to also play a role in the regulation of non-rapid eye movement (REM) sleep. Thus the immune response to infection may result in changes to the sleep cycle, including an increase in slow-wave sleep relative to REM sleep.

When suffering from sleep deprivation, active immunizations may have a diminished effect and may result in lower antibody production, and a lower immune response, than would be noted in a well-rested individual. Additionally, proteins such as NFIL3, which have been shown to be closely intertwined with both T-cell differentiation and our circadian rhythms, can be affected through the disturbance of natural light and dark cycles through instances of sleep deprivation, shift work, etc. As a result, these disruptions can lead to an increase in chronic conditions such as heart disease, chronic pain, and asthma.

In addition to the negative consequences of sleep deprivation, sleep and the intertwined circadian system have been shown to have strong regulatory effects on immunological functions affecting both the innate and the adaptive immunity. First, during the early slow-wave-sleep stage, a sudden drop in blood levels of cortisol, epinephrine, and norepinephrine induce increased blood levels of the hormones leptin, pituitary growth hormone, and prolactin. These signals induce a pro-inflammatory state through the production of the pro-inflammatory cytokines interleukin-1, interleukin-12, TNF-alpha and IFN-gamma. These cytokines then stimulate immune functions such as immune cells activation, proliferation, and differentiation. It is during this time that undifferentiated, or less differentiated, like naïve and central memory T cells, peak (i.e. during a time of a slowly evolving adaptive immune response). In addition to these effects, the milieu of hormones produced at this time (leptin, pituitary growth hormone, and prolactin) support the interactions between APCs and T-cells, a shift of the T_h1/T_h2 cytokine balance towards one that supports T_h1, an increase in overall T_h cell proliferation, and naïve T cell migration to lymph nodes. This milieu is also thought to support the formation of long-lasting immune memory through the initiation of Th1 immune responses.

In contrast, during wake periods differentiated effector cells, such as cytotoxic natural killer cells and CTLs (cytotoxic T lymphocytes), peak in order to elicit an effective response against any intruding pathogens. As well during awake active times, anti-inflammatory molecules, such as cortisol and catecholamines, peak. There are two theories as to why the pro-inflammatory state is reserved for sleep time. First, inflammation would cause serious cognitive and physical impairments if it were to occur during wake times. Second, inflammation may occur during sleep times due to the presence of melatonin. Inflammation causes a great deal of oxidative stress and the presence of melatonin during sleep times could actively counteract free radical production during this time.

Nutrition and Diet

Overnutrition is associated with diseases such as diabetes and obesity, which are known to affect immune function. More moderate malnutrition, as well as certain specific trace mineral and nutrient deficiencies, can also compromise the immune response.

Foods rich in certain fatty acids may foster a healthy immune system. Likewise, fetal undernourishment can cause a lifelong impairment of the immune system.

Eating vegetables and fruits can improve immune function.

Manipulation in Medicine

The immune response can be manipulated to suppress unwanted responses resulting from autoimmunity, allergy, and transplant rejection, and to stimulate protective responses against pathogens that largely elude the immune system or cancer.

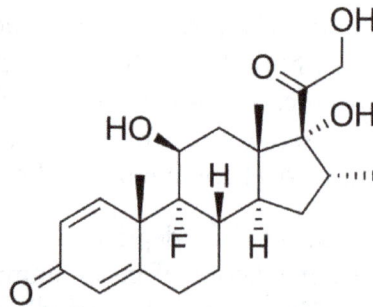

The immunosuppressive drug dexamethasone

Immunosuppression

Immunosuppressive drugs are used to control autoimmune disorders or inflammation when excessive tissue damage occurs, and to prevent transplant rejection after an organ transplant.

Anti-inflammatory drugs are often used to control the effects of inflammation. Glucocorticoids are the most powerful of these drugs; however, these drugs can have many undesirable side effects, such as central obesity, hyperglycemia, osteoporosis, and their use must be tightly controlled. Lower doses of anti-inflammatory drugs are often used in conjunction with cytotoxic or immunosuppressive drugs such as methotrexate or azathioprine. Cytotoxic drugs inhibit the immune response by killing dividing cells such as activated T cells. However, the killing is indiscriminate and other constantly dividing cells and their organs are affected, which causes toxic side effects. Immunosuppressive drugs such as cyclosporin prevent T cells from responding to signals correctly by inhibiting signal transduction pathways.

Immunostimulation

Cancer immunotherapy covers the medical ways to stimulate the immune system to attack cancer tumours.

Theoretical Approaches to The Immune System

Immunology is strongly experimental in everyday practice but is also characterized by an ongoing theoretical attitude. Many theories have been suggested in immunology from the end of the nineteenth century up to the present time. The end of the 19th century and the beginning of the 20th century saw a battle between "cellular" and "humoral" theories of immunity. According to the cellular theory of immunity, represented in particular by Elie Metchnikoff, it was cells – more precisely, phagocytes – that were responsible for immune responses. In contrast, the humoral theory of immunity, held, among others, by Robert Koch and Emil von Behring, stated that the active immune agents were soluble components (molecules) found in the organism's "humors" rather than its cells.

In the mid-1950s, Frank Burnet, inspired by a suggestion made by Niels Jerne, formulated the clonal selection theory (CST) of immunity. On the basis of CST, Burnet developed a theory of how an immune response is triggered according to the self/nonself distinction: "self" constituents

(constituents of the body) do not trigger destructive immune responses, while "nonself" entities (pathogens, an allograft) trigger a destructive immune response. The theory was later modified to reflect new discoveries regarding histocompatibility or the complex "two-signal" activation of T cells. The self/nonself theory of immunity and the self/nonself vocabulary have been criticized, but remain very influential.

More recently, several theoretical frameworks have been suggested in immunology, including "autopoietic" views, "cognitive immune" views, the "danger model" (or "danger theory", and the "discontinuity" theory. The danger model, suggested by Polly Matzinger and colleagues, has been very influential, arousing many comments and discussions.

Predicting Immunogenicity

Larger drugs (>500 Da) can provoke a neutralizing immune response, particularly if the drugs are administered repeatedly, or in larger doses. This limits the effectiveness of drugs based on larger peptides and proteins (which are typically larger than 6000 Da). In some cases, the drug itself is not immunogenic, but may be co-administered with an immunogenic compound, as is sometimes the case for Taxol. Computational methods have been developed to predict the immunogenicity of peptides and proteins, which are particularly useful in designing therapeutic antibodies, assessing likely virulence of mutations in viral coat particles, and validation of proposed peptide-based drug treatments. Early techniques relied mainly on the observation that hydrophilic amino acids are overrepresented in epitope regions than hydrophobic amino acids; however, more recent developments rely on machine learning techniques using databases of existing known epitopes, usually on well-studied virus proteins, as a training set. A publicly accessible database has been established for the cataloguing of epitopes from pathogens known to be recognizable by B cells. The emerging field of bioinformatics-based studies of immunogenicity is referred to as *immunoinformatics*. Immunoproteomics is the study of large sets of proteins (proteomics) involved in the immune response.

Manipulation by Pathogens

The success of any pathogen depends on its ability to elude host immune responses. Therefore, pathogens evolved several methods that allow them to successfully infect a host, while evading detection or destruction by the immune system. Bacteria often overcome physical barriers by secreting enzymes that digest the barrier, for example, by using a type II secretion system. Alternatively, using a type III secretion system, they may insert a hollow tube into the host cell, providing a direct route for proteins to move from the pathogen to the host. These proteins are often used to shut down host defenses.

An evasion strategy used by several pathogens to avoid the innate immune system is to hide within the cells of their host (also called intracellular pathogenesis). Here, a pathogen spends most of its life-cycle inside host cells, where it is shielded from direct contact with immune cells, antibodies and complement. Some examples of intracellular pathogens include viruses, the food poisoning bacterium *Salmonella* and the eukaryotic parasites that cause malaria (*Plasmodium falciparum*) and leishmaniasis (*Leishmania spp.*). Other bacteria, such as *Mycobacterium tuberculosis*, live inside a protective capsule that prevents lysis by complement. Many pathogens secrete compounds that diminish or misdirect the host's immune response.

Some bacteria form biofilms to protect themselves from the cells and proteins of the immune system. Such biofilms are present in many successful infections, e.g., the chronic *Pseudomonas aeruginosa* and *Burkholderia cenocepacia* infections characteristic of cystic fibrosis. Other bacteria generate surface proteins that bind to antibodies, rendering them ineffective; examples include *Streptococcus* (protein G), *Staphylococcus aureus* (protein A), and *Peptostreptococcus magnus* (protein L).

The mechanisms used to evade the adaptive immune system are more complicated. The simplest approach is to rapidly change non-essential epitopes (amino acids and/or sugars) on the surface of the pathogen, while keeping essential epitopes concealed. This is called antigenic variation. An example is HIV, which mutates rapidly, so the proteins on its viral envelope that are essential for entry into its host target cell are constantly changing. These frequent changes in antigens may explain the failures of vaccines directed at this virus. The parasite *Trypanosoma brucei* uses a similar strategy, constantly switching one type of surface protein for another, allowing it to stay one step ahead of the antibody response. Masking antigens with host molecules is another common strategy for avoiding detection by the immune system. In HIV, the envelope that covers the virion is formed from the outermost membrane of the host cell; such "self-cloaked" viruses make it difficult for the immune system to identify them as "non-self" structures.

Immunity (Medical)

In biology, immunity is the balanced state of having adequate biological defenses to fight infection, disease, or other unwanted biological invasion, while having adequate tolerance to avoid allergy, and autoimmune diseases.

Innate and Adaptive Immunity

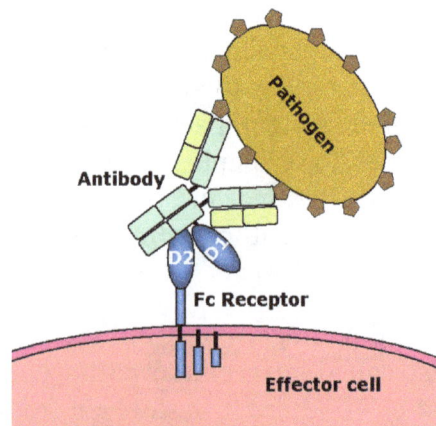

Scheme of a Fc receptor

It is the capability of the body to resist harmful microorganisms or viruses from entering it. Immunity involves both specific and nonspecific components. The nonspecific components act either as barriers or as eliminators of wide range of pathogens irrespective of antigenic specificity. Other components of the immune system adapt themselves to each new disease encountered and are able to generate pathogen-specific immunity.

The basic premise for the division of the immune system into innate and adaptive components comes down to the innate system being composed of primitive bone marrow cells that are programmed to recognise *foreign* substances and *react*, versus the adaptive system being composed of more advanced lymphatic cells that are programmed to recognise *self* substances and *don't react*. The reaction to foreign substances is etymologically described as inflammation, meaning *to set on fire*, while the non-reaction to self substances is etymologically described as immunity, meaning *to exempt*. The interaction of these two components of the immune system creates a dynamic biological environment where "Health" can be seen as an active physical state where what is self is immunologically spared, and what is foreign is inflammatorily and immunologically eliminated. Extending this concept, "Disease" then can arise when what is foreign cannot be eliminated, or what is self is not spared.

Innate immunity, or nonspecific immunity, is the natural resistances with which a person is born. It provides resistances through several physical, chemical and cellular approaches. Microbes first encounter the epithelial layers, physical barriers that line skin and mucous membranes. Subsequent general defences include secreted chemical signals (cytokines), antimicrobial substances, fever, and phagocytic activity associated with the inflammatory responses. The phagocytes express cell surface receptors that can bind and respond to common molecular patterns expressed on the surface of invading microbes. Through these approaches, innate immunity can prevent the colonization, entry and spread of microbes.

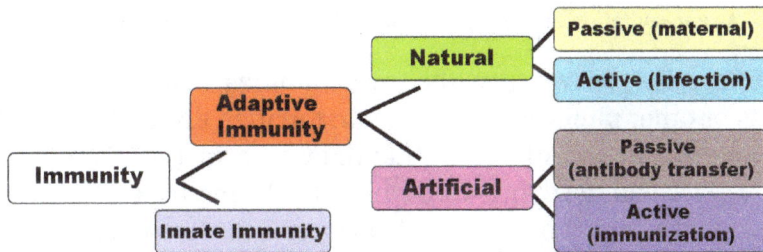

A further subdivision of adaptive immunity is characterized by the cells involved; humoral immunity is the aspect of immunity that is mediated by secreted antibodies, whereas the protection provided by cell mediated immunity involves T-lymphocytes alone. Humoral immunity is active when the organism generates its own antibodies, and passive when antibodies are transferred between individuals. Similarly, cell mediated immunity is active when the organisms' own T-cells are stimulated and passive when T cells come from another organism.

Adaptive immunity is often sub-divided into two major types depending on how the immunity was introduced. 'Naturally acquired immunity' occurs through contact with a disease causing agent, when the contact was not deliberate, whereas 'artificially acquired immunity' develops only through deliberate actions such as vaccination. Both naturally and artificially acquired immunity can be further subdivided depending on whether immunity is induced in the host or passively transferred from an immune host. 'Passive immunity' is acquired through transfer of antibodies or activated T-cells from an immune host, and is short lived—usually lasting only a few months—whereas 'active immunity' is induced in the host itself by antigen and lasts much longer, sometimes lifelong. The diagram below summarizes these divisions of immunity.

History of Theories of Immunity

The concept of immunity has intrigued mankind for thousands of years. The prehistoric view of disease was that it was caused by supernatural forces, and that illness was a form of theurgic punishment for "bad deeds" or "evil thoughts" visited upon the soul by the gods or by one's enemies. Between the time of Hippocrates and the 19th century, when the foundations of the scientific methods were laid, diseases were attributed to an alteration or imbalance in one of the four humors (blood, phlegm, yellow bile or black bile). Also popular during this time before learning that communicable diseases came from germs/microbes was the miasma theory, which held that diseases such as cholera or the Black Plague were caused by a miasma, a noxious form of "bad air". If someone were exposed to the miasma in a swamp, in evening air, or breathing air in a sickroom or hospital ward, they could get a disease.

A representation of the cholera epidemic of the nineteenth century.

The modern word "immunity" derives from the Latin *immunis*, meaning exemption from military service, tax payments or other public services. The first written descriptions of the concept of immunity may have been made by the Athenian Thucydides who, in 430 BC, described that when the plague hit Athens: "the sick and the dying were tended by the pitying care of those who had recovered, because they knew the course of the disease and were themselves free from apprehensions. For no one was ever attacked a second time, or not with a fatal result". The term "immunes", is also found in the epic poem "Pharsalia" written around 60 B.C. by the poet Marcus Annaeus Lucanus to describe a North African tribe's resistance to snake venom.

The first clinical description of immunity which arose from a specific disease causing organism is probably *Kitab fi al-jadari wa-al-hasbah* ('A Treatise on Smallpox and Measles', translated 1848) written by the Islamic physician Al-Razi in the 9th century. In the treatise, Al Razi describes the clinical presentation of smallpox and measles and goes on to indicate that that exposure to these specific agents confers lasting immunity (although he does not use this term). The first scientist who developed full theory of immunity was Ilya Mechnikov after he revealed phagocytosis in 1882. With Louis Pasteur's germ theory of disease, the fledgling science of immunology began to explain how bacteria caused disease, and how, following infection, the human body gained the ability to resist further infections.

The birth of active immunotherapy may have begun with Mithridates VI of Pontus. To induce active immunity for snake venom, he recommended using a method similar to modern toxoid serum therapy, by drinking the blood of animals which fed on venomous snakes. According to Jean de Maleissye, Mithridates assumed that animals feeding on venomous snakes acquired some detoxifying property in their bodies, and their blood must contain attenuated or transformed components of the snake venom. The action of those components might be strengthening the body to resist against the venom instead of exerting toxic effect. Mithridates reasoned that, by drinking the blood of these animals, he could acquire the similar resistance to the snake venom as the animals feeding on the snakes. Similarly, he sought to harden himself against poison, and took daily sub-lethal to build tolerance. Mithridates is also said to have fashioned a 'universal antidote' to protect him from all earthly poisons. For nearly 2000 years, poisons were thought to be the proximate cause of disease, and a complicated mixture of ingredients, called Mithridate, was used to cure poisoning during the Renaissance. An updated version of this cure, Theriacum Andromachi, was used well into the 19th century. In 1888 Emile Roux and Alexandre Yersin isolated diphtheria toxin, and following the 1890 discovery by Behring and Kitasato of antitoxin based immunity to diphtheria and tetanus, the antitoxin became the first major success of modern therapeutic Immunology.

Louis Pasteur in his laboratory, 1885.

In Europe, the induction of active immunity emerged in an attempt to contain smallpox. Immunization, however, had existed in various forms for at least a thousand years. The earliest use of immunization is unknown, however, around 1000 A.D. the Chinese began practicing a form of immunization by drying and inhaling powders derived from the crusts of smallpox lesions. Around the fifteenth century in India, the Ottoman Empire, and east Africa, the practice of inoculation (poking the skin with powdered material derived from smallpox crusts) became quite common. This practice was first introduced into the west in 1721 by Lady Mary Wortley Montagu. In 1798, Edward Jenner introduced the far safer method of deliberate infection with cowpox virus, (smallpox vaccine), which caused a mild infection that also induced immunity to smallpox. By 1800 the procedure was referred to as vaccination. To avoid confusion, smallpox inoculation was increasingly referred to as variolation, and it became common practice to use this term without regard for chronology. The success and general acceptance of Jenner's procedure would later drive the general nature of *vaccination* developed by Pasteur and others towards the end of the 19th century. In 1891, Pasteur widened the definition of vaccine in honour of Jenner and it then became essential to qualify the term, by referring to polio vaccine, measles vaccine etc.

Passive Immunity

Passive immunity is the transfer of active immunity, in the form of readymade antibodies, from one individual to another. Passive immunity can occur naturally, when maternal antibodies are transferred to the foetus through the placenta, and can also be induced artificially, when high levels of human (or horse) antibodies specific for a pathogen or toxin are transferred to non-immune individuals. Passive immunization is used when there is a high risk of infection and insufficient time for the body to develop its own immune response, or to reduce the symptoms of ongoing or immunosuppressive diseases. Passive immunity provides immediate protection, but the body does not develop memory, therefore the patient is at risk of being infected by the same pathogen later.

Naturally Acquired Passive Immunity

Maternal passive immunity is a type of naturally acquired passive immunity, and refers to antibody-mediated immunity conveyed to a fetus by its mother during pregnancy. Maternal antibodies (MatAb) are passed through the placenta to the fetus by an FcRn receptor on placental cells. This occurs around the third month of gestation. IgG is the only antibody isotype that can pass through the placenta. Passive immunity is also provided through the transfer of IgA antibodies found in breast milk that are transferred to the gut of the infant, protecting against bacterial infections, until the newborn can synthesize its own antibodies.

One of the first bottles of diphtheria antitoxin produced (Dated 1895).

Artificially Acquired Passive Immunity

Artificially acquired passive immunity is a short-term immunization induced by the transfer of antibodies, which can be administered in several forms; as human or animal blood plasma, as pooled human immunoglobulin for intravenous (IVIG) or intramuscular (IG) use, and in the form of monoclonal antibodies (MAb). Passive transfer is used prophylactically in the case of immunodeficiency diseases, such as hypogammaglobulinemia. It is also used in the treatment of several types of acute infection, and to treat poisoning. Immunity derived from passive immunization lasts for only a short period of time, and there is also a potential risk for hypersensitivity reactions, and serum sickness, especially from gamma globulin of non-human origin.

The artificial induction of passive immunity has been used for over a century to treat infectious disease, and prior to the advent of antibiotics, was often the only specific treatment for certain infections. Immunoglobulin therapy continued to be a first line therapy in the treatment of severe respiratory diseases until the 1930s, even after sulfonamide lot antibiotics were introduced.

Passive Transfer of Cell-Mediated Immunity

Passive or "adoptive transfer" of cell-mediated immunity, is conferred by the transfer of "sensitized" or activated T-cells from one individual into another. It is rarely used in humans because it requires histocompatible (matched) donors, which are often difficult to find. In unmatched donors this type of transfer carries severe risks of graft versus host disease. It has, however, been used to treat certain diseases including some types of cancer and immunodeficiency. This type of transfer differs from a bone marrow transplant, in which (undifferentiated) hematopoietic stem cells are transferred.

Active Immunity

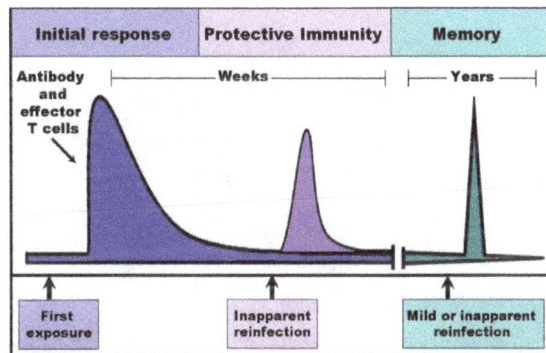

The time course of an immune response. Due to the formation of immunological memory, reinfection at later time points leads to a rapid increase in antibody production and effector T cell activity. These later infections can be mild or even unapparent.

When B cells and T cells are activated by a pathogen, memory B-cells and T- cells develop, and the *primary* immune response results. Throughout the lifetime of an animal these memory cells will "remember" each specific pathogen encountered, and are able to mount a strong *secondary* response, if the pathogen is detected again. The primary and secondary responses were first described in 1921 by English immunologist Alexander Glenny although the mechanism involved was not discovered until later.This type of immunity is both *active* and *adaptive* because the body's immune system prepares itself for future challenges. Active immunity often involves both the cell-mediated and humoral aspects of immunity as well as input from the innate immune system.

Naturally Acquired Active Immunity

Naturally acquired active immunity occurs when a person is exposed to a live pathogen, and develops a primary immune response, which leads to immunological memory. This type of immunity is "natural" because it is not induced by deliberate exposure. Many disorders of immune system function can affect the formation of active immunity such as immunodeficiency (both acquired and congenital forms) and immunosuppression.

Artificially Acquired Active Immunity

Artificially acquired active immunity can be induced by a vaccine, a substance that contains antigen. A vaccine stimulates a primary response against the antigen without causing symptoms of the disease. The term *vaccination* was coined by Richard Dunning, a colleague of Edward Jenner, and adapted by Louis Pasteur for his pioneering work in vaccination. The method Pasteur used entailed treating the infectious agents for those diseases so they lost the ability to cause serious disease. Pasteur adopted the name vaccine as a generic term in honor of Jenner's discovery, which Pasteur's work built upon.

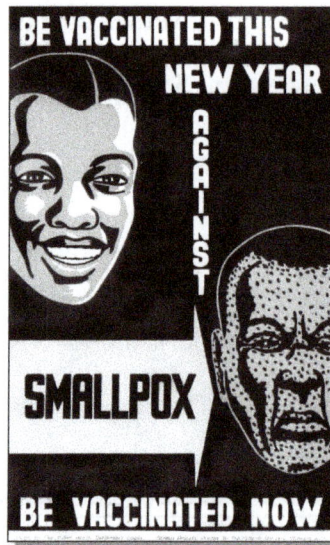

Poster from before the 1979 eradication of smallpox, promoting vaccination.

In 1807, Bavaria became the first group to require that their military recruits be vaccinated against smallpox, as the spread of smallpox was linked to combat. Subsequently the practice of vaccination would increase with the spread of war.

There are four types of traditional vaccines:

- Inactivated vaccines are composed of micro-organisms that have been killed with chemicals and/or heat and are no longer infectious. Examples are vaccines against flu, cholera, plague, and hepatitis A. Most vaccines of this type are likely to require booster shots.

- Live, attenuated vaccines are composed of micro-organisms that have been cultivated under conditions which disable their ability to induce disease. These responses are more durable and do not generally require booster shots. Examples include yellow fever, measles, rubella, and mumps.

- Toxoids are inactivated toxic compounds from micro-organisms in cases where these (rather than the micro-organism itself) cause illness, used prior to an encounter with the toxin of the micro-organism. Examples of toxoid-based vaccines include tetanus and diphtheria.

- Subunit vaccines are composed of small fragments of disease causing organisms. A characteristic example is the subunit vaccine against Hepatitis B virus.

Most vaccines are given by hypodermic or intramuscular injection as they are not absorbed reliably through the gut. Live attenuated polio and some typhoid and cholera vaccines are given orally in order to produce immunity based in the bowel.

Aspects of Immunity

Humoral Immunity

Humoral immunity, also called antibody-mediated immunity, is the aspect of immunity that is mediated by macromolecules (as opposed to cells) found in extracellular fluids such as secreted antibodies, complement proteins, and certain antimicrobial peptides. Humoral immunity is so named because it involves substances found in the humours, or body fluids.

The study of the molecular and cellular components that form the immune system, including their function and interaction, is the central science of immunology. The immune system is divided into a more primitive innate immune system, and acquired or adaptive immune system of vertebrates, each of which contains humoral and cellular components.

Humoral immunity refers to antibody production and the accessory processes that accompany it, including: Th2 activation and cytokine production, germinal center formation and isotype switching, affinity maturation and memory cell generation. It also refers to the effector functions of antibodies, which include pathogen and toxin neutralization, classical complement activation, and opsonin promotion of phagocytosis and pathogen elimination.

History

The concept of humoral immunity developed based on analysis of antibacterial activity of the serum components. Hans Buchner is credited with the development of the humoral theory. In 1890 he described alexins, or "protective substances", which exist in the blood serum and other bodily fluid and are capable of killing microorganisms. Alexins, later redefined "complement" by Paul Ehrlich, were shown to be the soluble components of the innate response that lead to a combination of cellular and humoral immunity, and bridged the features of innate and acquired immunity.

Following the 1888 discovery of the bacteria that cause diphtheria and tetanus, Emil von Behring and Kitasato Shibasaburō showed that disease need not be caused by microorganisms themselves. They discovered that cell-free filtrates were sufficient to cause disease. In 1890, filtrates of diphtheria, later named diphtheria toxins, were used to vaccinate animals in an attempt to demonstrate that immunized serum contained an antitoxin that could neutralize the activity of the toxin and could transfer immunity to non-immune animals. In 1897, Paul Ehrlich showed that antibodies form against the plant toxins ricin and abrin, and proposed that these antibodies are responsible for immunity. Ehrlich, with his friend Emil von Behring, went on to develop the diphtheria antitoxin, which became the first major success of modern immunotherapy. The presence and specificity of compatibility antibodies became the major tool for standardizing the state of immunity and identifying the presence of previous infections.

Major discoveries in the study of humoral immunity		
Substance	Activity	Discovery
Alexin(s) Complement	Soluble components in the serum that are capable of killing microorganisms	Buchner (1890), Ehrlich (1892)
Antitoxins	Substances in the serum that can neutralize the activity of toxins, enabling passive immunization	von Behring and Kitasato (1890)
Bacteriolysins	Serum substances that work with the complement proteins to induce bacterial lysis	Richard Pfeiffer (1895)
Bacterial agglutinins and precipitins	Serum substances that agglutinate bacteria and precipitate bacterial toxins	von Gruber and Durham (1896), Kraus (1897)
Hemolysins	Serum substances that work with complement to lyse red blood cells	Belfanti and Carbone (1898) Jules Bordet (1899)
Opsonins	Serum substances that coat the outer membrane of foreign substances and enhance the rate of phagocytosis by macrophages	Wright and Douglas (1903)
Antibody	Formation (1900), antigen-antibody binding hypothesis (1938), produced by B cells (1948), structure (1972), immunoglobulin genes (1976)	Founder: P Ehrlich

Complement System

The complement system is a biochemical cascade of the innate immune system that helps clear pathogens from an organism. It is derived from many small blood plasma proteins that work together to disrupt the target cell's plasma membrane leading to cytolysis of the cell. The complement system consists of more than 35 soluble and cell-bound proteins, 12 of which are directly involved in the complement pathways. The complement system is involved in the activities of both innate immunity and acquired immunity.

Activation of this system leads to cytolysis, chemotaxis, opsonization, immune clearance, and inflammation, as well as the marking of pathogens for phagocytosis. The proteins account for 5% of the serum globulin fraction. Most of these proteins circulate as zymogens, which are inactive until proteolytic cleavage.

Three biochemical pathways activate the complement system: the classical complement pathway, the alternate complement pathway, and the mannose-binding lectin pathway. The classical complement pathway typically requires antibodies for activation and is a specific immune response, while the alternate pathway can be activated without the presence of antibodies and is considered a non-specific immune response. Antibodies, in particular the IgG1 class, can also "fix" complement.

B cells

The principal function of B cells is to make antibodies against soluble antigens. B cell recognition of antigen is not the only element necessary for B cell activation (a combination of clonal proliferation and the terminal differentiation into plasma cells).

Naïve B cells can be activated in a T-cell dependent or independent manner, but two signals are always required to initiate activation.

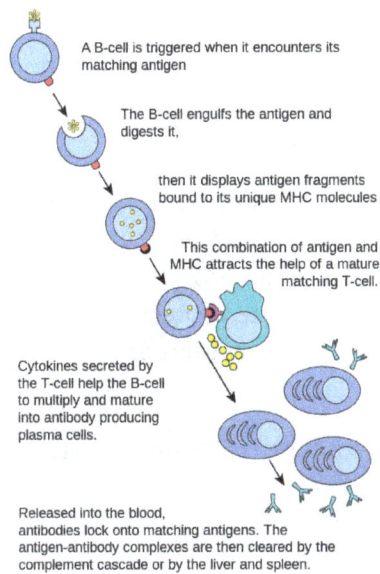

A B-cell is triggered when it encounters its matching antigen

The B-cell engulfs the antigen and digests it,

then it displays antigen fragments bound to its unique MHC molecules

This combination of antigen and MHC attracts the help of a mature matching T-cell.

Cytokines secreted by the T-cell help the B-cell to multiply and mature into antibody producing plasma cells.

Released into the blood, antibodies lock onto matching antigens. The antigen-antibody complexes are then cleared by the complement cascade or by the liver and spleen.

B cell activation is a large part of the humoral immune response.

B cell activation depends on one of three mechanisms: *Type 1 T cell-independent* (polyclonal) activation, *Type 2 T cell-independent* activation (in which mature B cells respond to highly repetitive structures causing cross-linking of the B cell receptors on the surface of B cells), and T cell-dependent activation. During T cell-dependent activation, an antigen presenting cell (APC) presents a processed antigen to a T helper cell (T_h), priming it. When a B cell processes and presents the *same* antigen to the *primed T_h cell*, the T cell releases cytokines that activate the B cell.

Antibodies

Immunoglobulins are glycoproteins in the immunoglobulin superfamily that function as antibodies. The terms *antibody* and *immunoglobulin* are often used interchangeably. They are found in the blood and tissue fluids, as well as many secretions. In structure, they are large Y-shaped globular proteins. In mammals there are five types of antibody: IgA, IgD, IgE, IgG, and IgM. Each immunoglobulin class differs in its biological properties and has evolved to deal with different antigens. Antibodies are synthesized and secreted by plasma cells that are derived from the B cells of the immune system.

An antibody is used by the acquired immune system to identify and neutralize foreign objects like bacteria and viruses. Each antibody recognizes a specific antigen unique to its target. By binding their specific antigens, antibodies can cause agglutination and precipitation of antibody-antigen products, prime for phagocytosis by macrophages and other cells, block viral receptors, and stimulate other immune responses, such as the complement pathway.

An incompatible blood transfusion causes a transfusion reaction, which is mediated by the humoral immune response. This type of reaction, called an acute hemolytic reaction, results in the rapid destruction (hemolysis) of the donor red blood cells by host antibodies. The cause is usually a clerical error, such as the wrong unit of blood being given to the wrong patient. The symptoms are fever and chills, sometimes with back pain and pink or red urine (hemoglobinuria). The major complication is that hemoglobin released by the destruction of red blood cells can cause acute renal failure.

Cell-Mediated Immunity

Cell-mediated immunity is an immune response that does not involve antibodies, but rather involves the activation of phagocytes, antigen-specific cytotoxic T-lymphocytes, and the release of various cytokines in response to an antigen.

Historically, the immune system was separated into two branches: humoral immunity, for which the protective function of immunization could be found in the humor (cell-free bodily fluid or serum) and cellular immunity, for which the protective function of immunization was associated with cells. CD4 cells or helper T cells provide protection against different pathogens. Cytotoxic T cells cause death by apoptosis without using cytokines, therefore in cell-mediated immunity cytokines are not *always* present.

The innate immune system and the adaptive immune system each comprise both humoral and cell-mediated components.

Cellular immunity protects the body by:

- T-cell mediated immunity or T-cell immunity : activating antigen-specific cytotoxic T cells that are able to induce apoptosis in body cells displaying epitopes of foreign antigen on their surface, such as virus-infected cells, cells with intracellular bacteria, and cancer cells displaying tumor antigens;

- activating macrophages and natural killer cells, enabling them to destroy pathogens; and

- stimulating cells to secrete a variety of cytokines that influence the function of other cells involved in adaptive immune responses and innate immune responses.

Cell-mediated immunity is directed primarily at microbes that survive in phagocytes and microbes that infect non-phagocytic cells. It is most effective in removing virus-infected cells, but also participates in defending against fungi, protozoans, cancers, and intracellular bacteria. It also plays a major role in transplant rejection.

References

- Goldsby RA; Kindt TK; Osborne BA & Kuby J (2003). Immunology (5th ed.). San Francisco: W.H. Freeman. ISBN 0-7167-4947-5.

- Smith A.D. (Ed) Oxford dictionary of biochemistry and molecular biology. (1997) Oxford University Press. ISBN 0-19-854768-4

- Alberts, Bruce; Alexander Johnson; Julian Lewis; Martin Raff; Keith Roberts; Peter Walters (2002). Molecular Biology of the Cell; Fourth Edition. New York and London: Garland Science. ISBN 978-0-8153-3218-3.

- Liszewski MK, Farries TC, Lublin DM, Rooney IA, Atkinson JP (1996). "Control of the complement system". Advances in Immunology. Advances in Immunology. 61: 201–83. doi:10.1016/S0065-2776(08)60868-8. ISBN 978-0-12-022461-6. PMID 8834497.

- Rajalingam R (2012). "Overview of the killer cell immunoglobulin-like receptor system". Methods in Molecular Biology. Methods in Molecular Biology™. 882: 391–414. doi:10.1007/978-1-61779-842-9_23. ISBN 978-1-61779-841-2.

- Holtmeier W, Kabelitz D (2005). "gammadelta T cells link innate and adaptive immune responses". Chemical Immunology and Allergy. Chemical Immunology and Allergy. 86: 151–83. doi:10.1159/000086659.

ISBN 3-8055-7862-8. PMID 15976493.

- Murphy, Kenneth; Weaver, Casey (2016). "10: The Humoral Immune Response". Immunobiology (9 ed.). Garland Science. ISBN 9780815345053.

- Romero P, Cerottini JC, Speiser DE (2006). "The human T cell response to melanoma antigens". Advances in Immunology. Advances in Immunology. 92: 187–224. doi:10.1016/S0065-2776(06)92005-7. ISBN 978-0-12-373636-9. PMID 17145305.

- Wira CR, Crane-Godreau M, Grant K (2004). "Endocrine regulation of the mucosal immune system in the female reproductive tract". In Ogra PL, Mestecky J, Lamm ME, Strober W, McGhee JR, Bienenstock J. Mucosal Immunology. San Francisco: Elsevier. ISBN 0-12-491543-4.

- Maleissye J (1991). Histoire Du Poison'.' Paris: Francois Bourin, ISBN 2-87686-082-1 (in French. Translated in Japanese: Hashimoto I, Katagiri T, translators (1996). [History of Poison]. Tokyo: Shin-Hyoron, Ltd., ISBN 4-7948-0315-X C0020).

- Janeway, Charles; Paul Travers; Mark Walport; Mark Shlomchik (2001). Immunobiology; Fifth Edition. New York and London: Garland Science. ISBN 0-8153-4101-6..

- Pradeu T; Jaeger S; Vivier E (2013). "The speed of change: towards a discontinuity theory of immunity?". Nature Reviews Immunology. 13 (10): 764–769. doi:10.1038/nri3521. PMID 23995627.

- Restifo, NP; Gattinoni, L (October 2013). "Lineage relationship of effector and memory T cells.". Current opinion in immunology. 25 (5): 556–63. doi:10.1016/j.coi.2013.09.003. PMID 24148236.

- Degn, S. E.; Thiel, S. (August 2013). "Humoral Pattern Recognition and the Complement System". Scandinavian Journal of Immunology. 78 (2): 181–193. doi:10.1111/sji.12070. PMID 23672641.

- Veldhoen, Marc; Brucklacher-Waldert, Verena (2012-10-01). "Dietary influences on intestinal immunity". Nature Reviews. Immunology. 12 (10): 696–708. doi:10.1038/nri3299. ISSN 1474-1741.

Branches of Immunology

The branches of immunology are based on the application of immunity research to various fields. This chapter informs the reader about branches like cancer immunology, reproductive immunology, testicular immunology, osteoimmunology, psychoneuroimmunology, immunoproteomics etc. The reader is provided with in-depth knowledge of each of these branches in this chapter.

Cancer Immunology

Cancer immunology is a branch of immunology that studies interactions between the immune system and cancer cells (also called tumors or malignancies). It is a field of research that aims to discover cancer immunotherapies to treat and retard progression of the disease. The immune response, including the recognition of cancer-specific antigens, forms the basis of targeted therapy (such as vaccines and antibody therapies) and tumor marker-based diagnostic tests. For instance tumour infiltrating lymphocytes are significant in human colorectal cancer. The host was given a better chance at survival if the cancer tissue showed infiltration of inflammatory cells, in particular those prompting lymphocytic reactions. The results yielded suggest some extent of anti-tumour immunity is present in colorectal cancers in humans.

Cancer immunosurveillance and immunoediting is based on (i) protection against development of spontaneous and chemically induced tumors in animal systems and (ii) identification of targets for immune recognition of human cancer.

Immunosurveillance

Cancer immunosurveillance is a theory formulated in 1957 by Burnet and Thomas, who proposed that lymphocytes act as sentinels in recognizing and eliminating continuously arising, nascent transformed cells. Cancer immunosurveillance appears to be an important host protection process that decreases cancer rates through inhibition of carcinogenesis and maintaining of regular cellular homeostasis. It has also been suggested that immunosurveillance primarily functions as a component of a more general process of cancer immunoediting.

Immunoediting

Immunoediting is a process by which a person is protected from cancer growth and the development of tumour immunogenicity by their immune system. It has three main phases: elimination, equilibrium and escape. The elimination phase consists of the following four phases:

Elimination

The first phase of elimination involves the initiation of an antitumor immune response. Cells of the innate immune system recognize the presence of a growing tumor which has undergone stromal remodeling, causing local tissue damage. This is followed by the induction of inflammatory signals which is essential for recruiting cells of the innate immune system (e.g. natural killer cells, natural killer T cells, macrophages and dendritic cells) to the tumor site. During this phase, the infiltrating lymphocytes such as the natural killer cells and natural killer T cells are stimulated to produce IFN-gamma.

In the second phase of elimination, newly synthesized IFN-gamma induces tumor death (to a limited amount) as well as promoting the production of chemokines CXCL10, CXCL9 and CXCL11. These chemokines play an important role in promoting tumor death by blocking the formation of new blood vessels. Tumor cell debris produced as a result of tumor death is then ingested by dendritic cells, followed by the migration of these dendritic cells to the draining lymph nodes. The recruitment of more immune cells also occurs and is mediated by the chemokines produced during the inflammatory process.

In the third phase, natural killer cells and macrophages transactivate one another via the reciprocal production of IFN-gamma and IL-12. This again promotes more tumor killing by these cells via apoptosis and the production of reactive oxygen and nitrogen intermediates. In the draining lymph nodes, tumor-specific dendritic cells trigger the differentiation of Th1 cells which in turn facilitates the development of cytotoxic CD8+ T cells also known as *killer T-cells*.

In the final phase of elimination, tumor-specific CD4+ and CD8+ T cells home to the tumor site and the cytotoxic T lymphocytes then destroy the antigen-bearing tumor cells which remain at the site.

Equilibrium and Escape

Tumor cell variants which have survived the elimination phase enter the equilibrium phase. In this phase, lymphocytes and IFN-gamma exert a selection pressure on tumor cells which are genetically unstable and rapidly mutating. Tumor cell variants which have acquired resistance to elimination then enter the escape phase. In this phase, tumor cells continue to grow and expand in an uncontrolled manner and may eventually lead to malignancies. In the study of cancer immunoediting, knockout mice have been used for experimentation since human testing is not possible. Tumor infiltration by lymphocytes is seen as a reflection of a tumor-related immune response.

Cancer Immunology and Chemotherapy

Obeid et al. investigated how inducing immunogenic cancer cell death ought to become a priority of cancer chemotherapy. He reasoned, the immune system would be able to play a factor via a 'bystander effect' in eradicating chemotherapy-resistant cancer cells. However, extensive research is still needed on how the immune response is triggered against dying tumour cells.

Professionals in the field have hypothesized that 'apoptotic cell death is poorly immunogenic whereas necrotic cell death is truly immunogenic'. This is perhaps because cancer cells being eradicated via a necrotic cell death pathway induce an immune response by triggering dendritic cells to mature, due to inflammatory response stimulation. On the other hand, apoptosis is connected to slight alterations within the plasma membrane causing the dying cells to be attractive to phagocytic cells. However, numerous animal studies have shown the superiority of vaccination with apoptotic cells, compared to necrotic cells, in eliciting anti-tumor immune responses.

Thus Obeid *et al.* propose that the way in which cancer cells die during chemotherapy is vital. Anthracyclins produce a beneficial immunogenic environment. The researchers report that when killing cancer cells with this agent uptake and presentation by antigen presenting dendritic cells is encouraged, thus allowing a T-cell response which can shrink tumours. Therefore activating tumour-killing T-cells is crucial for immunotherapy success.

However, advanced cancer patients with immunosuppression have left researchers in a dilemma as to how to activate their T-cells. The way the host dendritic cells react and uptake tumour antigens to present to CD4+ and CD8+ T-cells is the key to success of the treatment.

The Role of Viruses in Cancer Development

Various strains of human papillomavirus (HPV) have been found to play an important role in the development of cervical cancer. The HPV oncogenes E6 and E7 that these viruses possess have been shown to immortalise some human cells and thus promote cancer development. Although these strains of HPV have not been found in all cervical cancers, they have been found to be the cause in roughly 70% of cases. The study of these viruses and their role in the development of various cancers is still continuing, however a vaccine has been developed that can prevent infection of certain HPV strains, and thus prevent those HPV strains from causing cervical cancer, and possibly other cancers as well.

A virus that has been shown to cause breast cancer in mice is mouse mammary tumor virus. It is from discoveries such as this and the role of HPV in cervical cancer development that research is currently being undertaken to discover whether or not *human mammary tumour virus* is a cause of breast cancer in humans.

Reproductive Immunology

Reproductive immunology refers to a field of medicine that studies interactions (or the absence of them) between the immune system and components related to the reproductive system, such as maternal immune tolerance towards the fetus, or immunological interactions across the blood-testis barrier. The concept has been used by fertility clinics to explain the fertility problems, recurrent miscarriages and pregnancy complications observed when this state of immunological tolerance is not successfully achieved. Immunological therapy is the new up and coming method for treating many cases of previously "unexplained infertility" or recurrent miscarriage.

Between Mother and Fetus

The fact that the embryo's tissue is half foreign and unlike mismatched organ transplant, it is not normally rejected, suggests that the immunological system of the mother plays an important role in pregnancy. The placenta also plays an important part in protecting the embryo for the immune attack from the mother's system. Studies also propose that proteins in semen may help woman's immune system prepare for conception and pregnancy. For example, there is substantial evidence for exposure to partner's semen as prevention for pre-eclampsia, largely due to the absorption of several immune modulating factors present in seminal fluid, such as transforming growth factor beta (TGFβ).

Sperm Cells within A Male

The presence of anti-sperm antibodies in infertile men was first reported in 1954 by Rumke and Wilson. It has been noticed that the number of cases of sperm autoimmunity is higher in the infertile population leading to the idea that autoimmunity could be a cause of infertility.Anti sperm antigen has been described as three immunoglobulin isotopes (IgG, IgA, IgM) each of which targets different part of the spermatozoa. If more than 10% of the sperm are bound to anti-sperm antibodies (ASA), then infertility is suspected. The blood-testis barrier separates the immune system and the developing spermatozoa. The tight junction between the Sertoli cells form the blood-testis barrier but it is usually breached by physiological leakage. Not all sperms are protected by the barrier because spermatogonia and early spermatocytes are located below the junction. They are protected by other means like immunologic tolerance and immunomodulation.

Infertility after anti-sperm antibody binding can be caused by autoagglutination, sperm cytotoxicity, blockage of sperm-ovum interaction, and inadequate motility. Each presents itself depending on the binding site of ASA.

Immunocontraceptive Vaccine

Experiments are undergoing to test the effectiveness of an immunocontraceptive vaccine that inhibits the fusing of spermatozoa to the zona pellucida. This vaccine is currently being tested in animals and hopefully will be an effective contraceptive for humans. Normally, spermatozoa fuse with the zona pellucida surrounding the mature oocyte; the resulting acrosome reaction breaks down the egg's tough coating so that the sperm can fertilize the oovum. The mechanism of the vaccine is injection with cloned ZP cDNA, therefore this vaccine is a DNA based vaccine. This results in the production of antibodies against the ZP, which stop the sperm from binding to the zona pellucida and ultimately from fertilizing the oovum.

Another vaccine in investigation is one against HCG. This immunization would produce antibodies against hCG and TT. Antibodies against hCG would prevent the maintenance of the uterus for a viable pregnancy therefore preventing contraception. Another vaccine that is utilized is the peptide β-hCG that is more specific to hCG and a more rapid and effective response occurs in the absence of LH, FSH, and TSH.

Testicular Immunology

Testicular Immunology is the study of the immune system within the testis. It includes an investigation of the effects of infection, inflammation and immune factors on testicular function. Two unique characteristics of testicular immunology are evident: (1) the testis is described as an immunologically privileged site, where suppression of immune responses occurs; and, (2) some factors which normally lead to inflammation are present at high levels in the testis, where they regulate the development of sperm instead of promoting inflammation.

History of Testicular Immunology

- 460-377 BC Hippocrates described testicular inflammation associated with mumps

- 1785 Hunter and Michaelis performed transplant experiments in domestic chickens

- 1849 Berthold transplanted testes between roosters and showed maintenance of male sex characteristics only in birds with successfully grafted testes

- 1899-1900 Sperm recognized as immunogenic (will cause an autoimmune reaction if transplanted from the testis into a different area of the body) by Landsteiner (1899) and Metchinikoff, (1900)

- 1913-1914 Human testis transplants performed by Lespinasse (1913), and Lydson (1914) who performed a graft on himself!

- 1954 Discovery that sperm autoantibodies contribute to infertility,

- 1977 Billingham recognized that the testis is site of immune privilege

Immune Cells Found in The Testis

Immune cells of the human testis are not as well characterized as those from rodents, due to the rarity of normal human testes available for experiment. The majority of experiments have studied the rat testis due to its convenience: it is of relatively large size and is easily extracted from experimental animals.

Macrophages

Macrophages are directly involved in the fight against invading micro-organisms as well as being antigen-presenting cells which activate lymphocytes. Early studies demonstrated the presence of macrophages in the rat testis Testicular macrophages are the largest population of immune cells in the rodent testis. Macrophages have also been found in the testes of humans, guinea pigs, hamsters, boars, horses and bulls. They originate from blood monocytes which move into the testis then mature into macrophages. In the rat, testicular macrophages have been described as either "resident" or "newly arrived" from the blood supply. It is likely that most of the adult population of testicular macrophages in adult rats are a result of very rapid proliferation of early precursors that entered the testis during postnatal maturation

Testicular macrophages can respond to infectious stimuli and become activated (undergo changes enabling the killing of the invading micro-organism), but do so to a lesser extent than other types of macrophages. An example is production of the inflammatory cytokines TNFα and IL-1β by activated rat testicular macrophages: these macrophages produce significantly less TNFα and IL-1β than activated rat peritoneal macrophages. Aside from responding to infectious stimuli, testicular macrophages are also involved in maintaining normal testis function. They have been shown to secrete 25-hydroxycholesterol, a sterol that can be converted to testosterone by Leydig cells. Their presence is necessary for the normal development and function of the Leydig cells, which are the testosterone-producing cells of the testis.

B-Lymphocytes

B-lymphocytes take part in the adaptive immune response and produce antibodies. These cells are not normally found in the testis, even during inflammatory conditions. The lack of B-lymphocytes in the testis is significant, since these are the antibody-producing cells of the immune system. Since anti-sperm antibodies can cause infertility, it is important that antibody-producing B-lymphocytes are kept separated from the testis.

T-lymphocytes

T-lymphocytes (T-cells) are white blood cells which take part in cell-mediated immunity. They are often found within tissues where they can be activated by antigen-presenting cells upon infection. They are present in rat and human testes, where they constitute approximately 10 to 20% of the immune cells present, as well as mouse and ram testes. Both cytotoxic T-cells and Helper T cells are found in the testes of rats. Also present in the testes of rats and humans are natural killer cells and Natural killer T cells have been found in rats and mice.

Mast cells

Mast cells are regulators of immune responses, particularly those against parasites. They are also involved in the development of autoimmune diseases and allergies. Mast cells have been found in relatively low numbers in the testes of humans, rats, mice, dogs, cats, bulls, boars and deer. In the mammalian testis mast cells regulate testosterone production. There are two lines of evidence that restriction of mast cell activation in the testis could be beneficial during treatment of inflammatory conditions; (1) In experimental models of testicular inflammation, mast cells were present in 10-fold greater numbers and showed signs of activation, and (2) Treatment with drugs which stabilize mast cell activation has proved beneficial in treating some types of male infertility.

Eosinophils

Eosinophils directly fight parasitic infections and are involved in allergic reactions. They have been found in relatively low numbers in the rat, mouse, dog, cat, bull and deer testes. Almost nothing is known about their significance or function in the testis.

Dendritic cells

Dendritic cells initiate adaptive immune responses. Relatively small amounts of dendritic cells have been found in the testes of humans, rats and mice. The functional role of dendritic cells in the testis is not well understood, although they have been shown to be involved in autoimmune orchitis during animal experiments. When autoimmune orchitis is induced in rats, the dendritic cell population of the testis greatly increases. This is likely to contribute to testicular inflammation, considering the well-established role of dendritic cells in other types of autoimmune inflammation.

Neutrophils

Neutrophils are white blood cells which are present in the blood but not normally in tissues. They move out from the blood into tissues and organs upon infection or damage. They directly fight invading pathogens such as bacteria. Neutrophils are not found in the rodent testis under normal conditions but can enter from the blood supply upon infection or inflammatory stimulus. This has been demonstrated in the rat after injection with bacterial cell wall components to produce an immune reaction. Neutrophils also enter the rat testis after treatment with hormones that increase the permeability of blood vessels. In humans, neutrophils have been found in the testis when associated with some tumors. In rat experiments, testicular torsion leads to neutrophil entry into the testis. Neutrophil activity in the testis is an inflammatory response which needs to be tightly regulated by the body, since inflammation-induced damage to the testis can lead to infertility. It is assumed that the role of the immunosuppressive environment of the testis is to protect developing sperm from inflammation.

Immune Privilege in The Testis

Sperm are immunogenic - that is they will cause an autoimmune reaction if transplanted from the testis into a different part of the body. This has been demonstrated in experiments using rats by Landsteiner (1899) and Metchinikoff (1900), mice and guinea pigs. The likely reason for this is that sperm first mature at puberty, after immune tolerance is established, therefore the body recognizes them as foreign and mounts an immune reaction against them. Therefore, mechanisms for their protection must exist in this organ to prevent any autoimmune reaction. The blood-testis barrier is likely to contribute to the survival of sperm. However, it is believed in the field of testicular immunology that the blood-testis barrier cannot account for all immune suppression in the testis, due to (1) its incompleteness at a region called the rete testis and (2) the presence of immunogenic molecules outside the blood-testis barrier, on the surface of spermatogonia. Another mechanism which is likely to protect sperm is the suppression of immune responses in the testis. Both the suppression of immune responses and the increased survival of grafts in the testis have led to its recognition as an immunologically privileged site. Other immunologically privileged sites include the eye, brain and uterus.

The two main features of immune privilege in the rat testis are;

- a diminishment in the activation of testicular macrophages by infections such as bacteria, and

- a defect in the activation of T-cells when antigen is presented to them, leading to the absence of an adaptive immune response to sperm in the testis.

It is also predicted that the high level of inflammatory cytokines in the testis contributes to immune privilege.

Immune Privilege in Rodents and Other Experimental Animals

The existence of immune privilege in the testes of rodents is well accepted, due to many experiments demonstrating prolonged, and sometimes indefinite, survival of tissue transplanted into the testis, or testicular tissue transplanted elsewhere. Evidence includes the tolerance of testicular grafts in mice and rats, as well as the increased survival of transplants of pancreatic insulin-producing cells in rats, when cells from the testes (Sertoli cells) are added to the transplanted material. Complete spermatogenesis, forming functional pig or goat sperm, can be established by the grafting of pig or goat testicular tissue onto the backs of mice - however, immunodeficient mice needed to be used.

Immune Privilege in Humans

The presence of immune-privilege in the human testis is controversial and insufficient evidence exists to either confirm or rule out this phenomenon.

- Evidence for human/primate testicular immune privilege:

Sperm are protected from autoimmune attack, which when it occurs in humans leads to infertility. Local injury of seminiferous tubules caused by fine-needle biopsies in humans does not cause testicular inflammation (orchitis). Furthermore, human testis cells tolerate early HIV infection with little response.

- Evidence against human/primate testicular immune-privilege:

In transplant experiments, primate testes fail to support grafts of monkey thyroid tissue. Human testis tissue transplanted into the mouse elicited an immune response and was rejected, however, this immune response was not as extensive as that against other types of grafted tissue.

How Does The Testis Suppress Immune Responses?

How the testicular environment suppresses the immune response is only partially understood. Recent experiments have uncovered a number of biological processes that most likely contribute to immune privilege in the testes of rodents:

- Experiments in the rat have shown that Sertoli cells can help protect from graft rejection. These cells were isolated from the testis, then added to transplants of the insulin-producing cells of the pancreas (islets of Langerhans), resulting in increased graft survival. Molecules released by the Sertoli cells are predicted to protect the graft.

- It is likely that the testicular environment itself is inhibiting the activation of T-cells, in order to protect the developing sperm which are immunogenic. The fluid present in the testis is a potent inhibitor of the activation of T-cells under laboratory conditions.

- The diminishment of the testis inflammatory response is likely to result from relatively low levels of inflammatory cytokines released by activated testicular macrophages.

Since protection of developing sperm is so important to the survival of a species, it would not be surprising if more than one mechanism were in use.

Immune Factors Regulate Normal Testis Function

Curiously, the testis contains factors such as cytokines, which are usually only produced upon infections and tissue damage. The cytokines interleukin-1α (IL-1α), IL-6 and Activin A are found in the testis, often at high levels. In other tissues, these cytokine would promote inflammation, but here they control testis function. They regulate the development of sperm by controlling their cell division and survival.

Other immune factors found in the testis include the enzyme inducible nitric oxide synthase (iNOS), and its product nitric oxide (NO), transforming growth factor beta (TGFβ), the enzyme cyclooxygenase-2 (COX-2) and its product prostaglandin E2, and many others. Further research is required to define the functional roles of these immune factors in the testis.

The Effects of Infections and Immune Responses on The Testis

Mumps

Mumps is a viral disease which causes swelling of the salivary glands and testes. The mumps virus lives in the upper respiratory tract and spreads through direct contact with saliva. Prior to widespread vaccination programs, it was a common childhood disease. Mumps is generally not serious in children, but in adults, where sperm have matured in the testis, it can cause more severe complications, such as infertility.

Sexually Transmitted Diseases

Gonorrhea is a sexually transmitted disease caused by the bacteria *Niesseria gonorrhea* which can lead to testicular pain and swelling. Gonorrhea also infects the female reproductive system around the cervix and uterus, and can grow in the mouth, throat, eyes and anus. It can be effectively treated with antibiotics, however, if untreated, gonorrhea can cause infertility in men. Chlamydia is caused by the sexually transmitted bacteria *Chlamydia trachomatis* which infects the genitals. It more commonly affects women, and if untreated, can lead to pelvic inflammatory disease and infertility. Serious symptoms in men are rare, but include swollen testicles and an unusual discharge from the penis. It is effectively treated with antibiotics.

Anti-Sperm Antibodies

Anti-sperm antibodies are produced by cells of the adaptive immune system and are the cause of approximately 7% of male infertility incidences. They can bind to sperm, inhibiting their movement, stopping recognition and entry into the egg, or targeting sperm for destruction when they reach the female reproductive tract

Testicular Torsion

Testicular torsion is a condition of physical twisting of the testis which results in cutting off the blood supply. It leads to damage that, if not treated within a few hours, causes the death of testicular tissue, and requires removal of the testis to prevent gangrene, and therefore can cause infertility.

Autoimmune Orchitis

Orchitis is a condition of testicular pain involving swelling, inflammation and possibly infection. Orchitis can be caused by an autoimmune reaction (autoimmune orchitis) leading to a reduction in fertility. Autoimmune orchitis is rare in humans, compared to anti-sperm antibodies. To study orchitis in the testis, autoimmune orchitis has been induced in the rodent testis. The disease starts with the appearance of testicular antibodies, then movement of macrophages and lymphocytes from the blood stream into the testis, breaking of the physical interactions between the developing sperm and Sertoli cells, entry of neutrophils or eosinophils, and finally death of the developing sperm, leading to infertility.

Inflammation Models in The Rodent

Experiments in rats have examined, in fine detail, the course of testicular events during a bacterial infection. In the short term (3 hours) multiple inflammatory factors are produced and released by testicular macrophages. Examples are prostaglandin E2, inducible nitric oxide synthase (iNOS), TNFα and IL-1β, although at lower levels than other tissues. Non-immune cells of the testis such as Sertoli cells and Leydig cells also able to respond to bacteria. During a bacterial infection, testosterone levels and the amount of testicular interstitial fluid are reduced. Neutrophils enter the testis about 12 hours after infection. Importantly, there is damage to the developing sperm, which start to die under severe infections. Despite all the data on the effects of bacteria on normal testis parameters, there is little experimental data regarding its effect on rodent fertility.

Other Diseases Where Testicular Inflammation can be A Symptom

Testicular inflammation can be a symptom of the following diseases: Coxsackie A virus, varicella (chicken pox) human immunodeficiency virus (HIV), dengue fever, Epstein Barr virus-associated infectious mononucleosis, syphilis, leprosy, tuberculosis.

Immunoproteomics

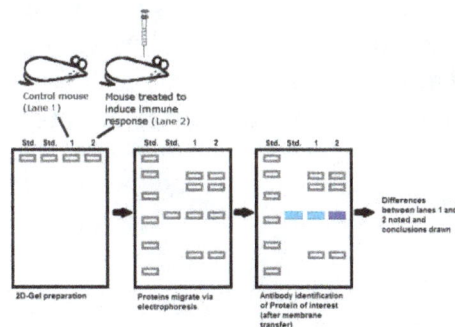

Immunoproteomics example experiment involving western blot analysis

Immunoproteomics is the study of large sets of proteins (proteomics) involved in the immune response.

Examples of common applications of immunoproteomics include:

- The isolation and mass spectrometric identification of MHC (major histocompatibility complex) binding peptides

- Purification and identification of protein antigens binding specific antibodies (or other affinity reagents)

- Comparative immunoproteomics to identify proteins and pathways modulated by a specific infectious organism, disease or toxin.

The identification of proteins in immunoproteomics is carried out by techniques including gel based, microarray based, and DNA based techniques, with mass spectroscopy typically being the ultimate identification method.

Applications

Immunology

Immunoproteomics is and has been used to increase scientific understanding of both autoimmune disease pathology and progression. Using biochemical techniques, gene and ultimately protein expression can be measured with high fidelity. With this information, the biochemical pathways causing pathology in conditions such as multiple sclerosis and Crohn's disease can potentially be elucidated. Serum antibody identification in particular has proven to be very useful as a diagnostic tool for a number of diseases in modern medicine, in large part due to the relatively high stability of serum antibodies.

Immunoproteomic techniques are additionally used for the isolation of antibodies. By identifying and proceeding to sequence antibodies, scientists are able to identify potential protein targets of said antibodies. In doing so, it is possible to determine the antigen(s) responsible for a particular immune response. Identification and engineering of antibodies involved in autoimmune disease pathology may offer novel techniques in disease therapy.

Drug Engineering

By identifying the antigens responsible for a particular immune response, it is possible to identify viable targets for novel drugs. In addition, specific antigens can further be classified based on immunoreactivity for identification of future potential vaccine preparations. In addition to the identification of vaccine candidates, immunoproteomic techniques such as western blotting can additionally be used for measuring the efficacy of a given vaccine.

Technology and Instrumentation

Mass Spectrometry

Mass spectrometry can be used in the sequencing of MHC binding motifs, which can subsequently be used to predict T cell epitopes. The technique of peptide mass fingerprinting (PMF) can be used to check a peptide's mass spectrum against a database of protein digests which have already been documented. If the mass spectrum of the protein of interest as well as the database protein share a large amount of homology, it is likely that the protein of interest is contained within the sample.

2-D Gel Electrophoresis and Western Blotting

Two-dimensional gel electrophoresis (2-D gel) techniques in culmination with western blotting has been used for many years in the identification of immune response magnitude. This can be accomplished by comparing various samples against molecular-weight size markers for qualitative analysis and against known amounts of protein standards for quantitative analysis.

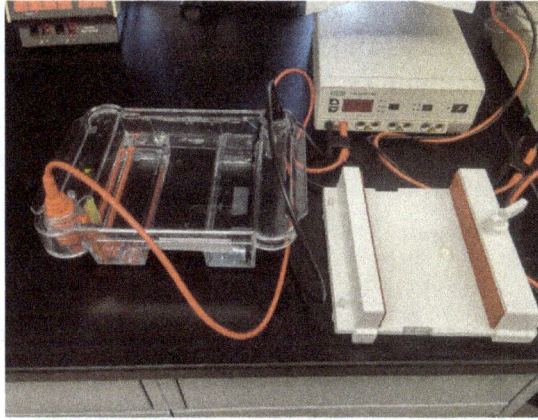

Common 2D-gel electrophoresis apparatus

2-D Liquid Chromatography

By coupling liquid chromatography with a variety of other immunodetection techniques such as serological proteome analysis (SERPA), it is possible to analyze the hydrophobicity, PI, relative mass, and antibody reactivity of antibodies within a given serum.

Microarray

Microarray analysis of various serums can be used as a means to identify changes in gene expression before, after, and during a given immune response.

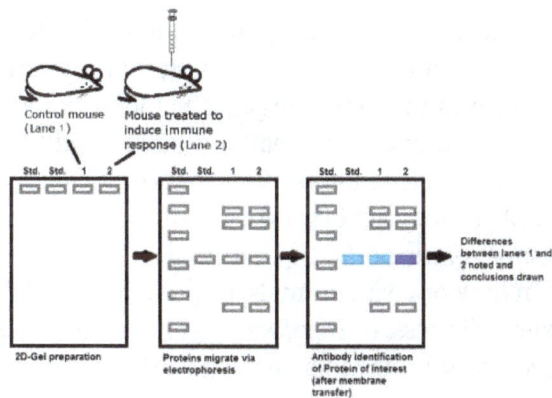

Immunoproteomic mice example

Psychoneuroimmunology

Psychoneuroimmunology (PNI), also referred to as psychoendoneuroimmunology (PENI) or psychoneuroendocrinoimmunology (PNEI), is the study of the interaction between psychological processes and the nervous and immune systems of the human body. PNI takes an interdisciplinary approach, incorporating psychology, neuroscience, immunology, physiology, genetics, pharmacology, molecular biology, psychiatry, behavioral medicine, infectious diseases, endocrinology, and rheumatology.

The main interests of PNI are the interactions between the nervous and immune systems and the relationships between mental processes and health. PNI studies, among other things, the physiological functioning of the neuroimmune system in health and disease; disorders of the neuroimmune system (autoimmune diseases; hypersensitivities; immune deficiency); and the physical, chemical and physiological characteristics of the components of the neuroimmune system in vitro, in situ, and in vivo.

History

Interest in the relationship between psychiatric syndromes or symptoms and immune function has been a consistent theme since the beginning of modern medicine.

Claude Bernard, the father of modern physiology, with his pupils

Claude Bernard, a French physiologist of the Muséum national d'Histoire naturelle, formulated the concept of the *milieu interieur* in the mid-1800s. In 1865, Bernard described the perturbation of this internal state: "... there are protective functions of organic elements holding living materials in reserve and maintaining without interruption humidity, heat and other conditions indispensable to vital activity. Sickness and death are only a dislocation or perturbation of that mechanism" (Bernard, 1865). Walter Cannon, a professor of physiology at Harvard University coined the commonly used term, homeostasis, in his book *The Wisdom of the Body*, 1932, from the Greek word *homoios*, meaning similar, and *stasis*, meaning position. In his work with animals, Cannon observed that any change of emotional state in the beast, such as anxiety, distress, or rage, was accompanied by total cessation of movements of the stomach (*Bodily Changes in Pain, Hunger, Fear and Rage*, 1915). These studies into the relationship between the effects of emotions and perceptions on the autonomic nervous system, namely the sympathetic and parasympathetic responses that initiated the recognition of the freeze, fight or flight response. His findings were published from time to time in professional journals, then summed up in book form in *The Mechanical Factors of Digestion*, published in 1911.

Hans Selye, a student of Johns Hopkins University and McGill University, and a researcher at Université de Montréal, experimented with animals by putting them under different physical and mental adverse conditions and noted that under these difficult conditions the body consistently adapted to heal and recover. Several years of experimentation that formed the empiric foundation of Selye's concept of the General Adaptation Syndrome. This syndrome consists of an enlargement of the adrenal gland, atrophy of the thymus, spleen, and other lymphoid tissue, and gastric ulcerations.

Bust of Hans Selye at Selye János University, Komárno, Slovakia

Selye describes three stages of adaptation, including an initial brief alarm reaction, followed by a prolonged period of resistance, and a terminal stage of exhaustion and death. This foundational work led to a rich line of research on the biological functioning of glucocorticoids.

Mid-20th century studies of psychiatric patients reported immune alterations in psychotic individuals, including lower numbers of lymphocytes and poorer antibody response to pertussis vaccination, compared with nonpsychiatric control subjects. In 1964, George F. Solomon, from the University of California in Los Angeles, and his research team coined the term "psychoimmunology" and published a landmark paper: "Emotions, immunity, and disease: a speculative theoretical integration."

Origins

In 1975, Robert Ader and Nicholas Cohen, at the University of Rochester, advanced PNI with their demonstration of classic conditioning of immune function, and they subsequently coined the term "psychoneuroimmunology". Ader was investigating how long conditioned responses (in the sense of Pavlov's conditioning of dogs to drool when they heard a bell ring) might last in laboratory rats. To condition the rats, he used a combination of saccharin-laced water (the conditioned stimulus) and the drug Cytoxan, which unconditionally induces nausea and taste aversion and suppression of immune function. Ader was surprised to discover that after conditioning, just feeding the rats saccharin-laced water was associated with the death of some animals and he proposed that they had been immunosuppressed after receiving the conditioned stimulus. Ader (a psychologist) and Cohen (an immunologist) directly tested this hypothesis by deliberately immunizing conditioned and unconditioned animals, exposing these and other control groups to the conditioned taste stimulus, and then measuring the amount of antibody produced. The highly reproducible results revealed that conditioned rats exposed to the conditioned stimulus were indeed immuno suppressed. In other words, a signal via the nervous system (taste) was affecting immune function. This was one of the first scientific experiments that demonstrated that the nervous system can affect the immune system.

In 1981, David L. Felten, then working at the Indiana University School of Medicine, discovered a network of nerves leading to blood vessels as well as cells of the immune system. The researcher, along with his team, also found nerves in the thymus and spleen terminating near clusters of lymphocytes, macrophages, and mast cells, all of which help control immune function. This discovery provided one of the first indications of how neuro-immune interaction occurs.

Ader, Cohen, and Felten went on to edit the groundbreaking book *Psychoneuroimmunology* in 1981, which laid out the underlying premise that the brain and immune system represent a single, integrated system of defense.

In 1985, research by neuropharmacologist Candace Pert, of the National Institutes of Health at Georgetown University, revealed that neuropeptide-specific receptors are present on the cell walls of both the brain and the immune system. The discovery that neuropeptides and neurotransmitters act directly upon the immune system shows their close association with emotions and suggests mechanisms through which emotions, from the limbic system, and immunology are deeply interdependent. Showing that the immune and endocrine systems are modulated not only by the brain but also by the central nervous system itself affected the understanding of emotions, as well as disease.

Contemporary advances in psychiatry, immunology, neurology, and other integrated disciplines of medicine has fostered enormous growth for PNI. The mechanisms underlying behaviorally induced alterations of immune function, and immune alterations inducing behavioral changes, are likely to have clinical and therapeutic implications that will not be fully appreciated until more is known about the extent of these interrelationships in normal and pathophysiological states.

The Immune-Brain Loop

PNI research is looking for the exact mechanisms by which specific brainimmunity effects are achieved. Evidence for nervous system–immune system interactions exists at several biological levels.

The immune system and the brain talk to each other through signaling pathways. The brain and the immune system are the two major adaptive systems of the body. Two major pathways are involved in this cross-talk: the Hypothalamic-pituitary-adrenal axis (HPA axis) and the sympathetic nervous system (SNS). The activation of SNS during an immune response might be aimed to localize the inflammatory response.

The body's primary stress management system is the HPA axis. The HPA axis responds to physical and mental challenge to maintain homeostasis in part by controlling the body's cortisol level. Dysregulation of the HPA axis is implicated in numerous stress-related diseases, with evidence from meta-analyses indicating that different types/duration of stressors and unique personal variables can shape the HPA response. HPA axis activity and cytokines are intrinsically intertwined: inflammatory cytokines stimulate adrenocorticotropic hormone (ACTH) and cortisol secretion, while, in turn, glucocorticoids suppress the synthesis of proinflammatory cytokines.

Molecules called pro-inflammatory cytokines, which include interleukin-1 (IL-1), Interleukin-2 (IL-2), interleukin-6 (IL-6), Interleukin-12 (IL-12), Interferon-gamma (IFN-Gamma) and tumor necrosis factor alpha (TNF-alpha) can affect brain growth as well as neuronal function. Circulating immune cells such as macrophages, as well as glial cells (microglia and astrocytes) secrete these molecules. Cytokine regulation of hypothalamic function is an active area of research for the treatment of anxiety-related disorders.

Cytokines mediate and control immune and inflammatory responses. Complex interactions exist between cytokines, inflammation and the adaptive responses in maintaining homeostasis. Like the stress response, the inflammatory reaction is crucial for survival. Systemic inflammatory reaction results in stimulation of four major programs:

- the acute-phase reaction
- sickness behavior
- the pain program
- the stress response

These are mediated by the HPA axis and the SNS. Common human diseases such as allergy, autoimmunity, chronic infections and sepsis are characterized by a dysregulation of the pro-inflammatory versus anti-inflammatory and T helper (Th1) versus (Th2) cytokine balance.

Recent studies show pro-inflammatory cytokine processes take place during depression, mania and bipolar disease, in addition to autoimmune hypersensitivity and chronic infections.

Chronic secretion of stress hormones, glucocorticoids (GCs) and catecholamines (CAs), as a result of disease, may reduce the effect of neurotransmitters, including serotonin, norepinephrine and dopamine, or other receptors in the brain, thereby leading to the dysregulation of neurohormones. Under stimulation, norepinephrine is released from the sympathetic nerve terminals in organs, and the target immune cells express adrenoreceptors. Through stimulation of these receptors, locally released norepinephrine, or circulating catecholamines such as epinephrine, affect lymphocyte traffic, circulation, and proliferation, and modulate cytokine production and the functional activity of different lymphoid cells.

Glucocorticoids also inhibit the further secretion of corticotropin-releasing hormone from the hypothalamus and ACTH from the pituitary (negative feedback). Under certain conditions stress hormones may facilitate inflammation through induction of signaling pathways and through activation of the Corticotropin-releasing hormone.

These abnormalities and the failure of the adaptive systems to resolve inflammation affect the well-being of the individual, including behavioral parameters, quality of life and sleep, as well as indices of metabolic and cardiovascular health, developing into a "systemic anti-inflammatory feedback" and/or "hyperactivity" of the local pro-inflammatory factors which may contribute to the pathogenesis of disease.

This systemic or neuro-inflammation and neuroimmune activation have been shown to play a role in the etiology of a variety of neurodegenerative disorders such as Parkinson's and Alzheimer's disease, multiple sclerosis, pain, and AIDS-associated dementia. However, cytokines and chemokines also modulate central nervous system (CNS) function in the absence of overt immunological, physiological, or psychological challenges.

Psychoneuroimmunological Effects

There is now sufficient data to conclude that immune modulation by psychosocial stressors and/or interventions can lead to actual health changes. Although changes related to infectious disease and wound healing have provided the strongest evidence to date, the clinical importance of immunological dysregulation is highlighted by increased risks across diverse conditions and diseases. For example, stressors can produce profound health consequences. In one epidemiological study, all-cause mortality increased in the month following a severe stressor – the death of a spouse. Theorists propose that stressful events trigger cognitive and affective responses which, in turn, induce sympathetic nervous system and endocrine changes, and these ultimately impair immune function. Potential health consequences are broad, but include rates of infection HIV progression cancer incidence and progression, and high rates of infant mortality.

Understanding Stress and Immune Function

Stress is thought to affect immune function through emotional and/or behavioral manifestations such as anxiety, fear, tension, anger and sadness and physiological changes such as heart rate, blood pressure, and sweating. Researchers have suggested that these changes are beneficial if they are of limited duration, but when stress is chronic, the system is unable to maintain equilibrium or homeostasis.

In one of the earlier PNI studies, which was published in 1960, subjects were led to believe that they had accidentally caused serious injury to a companion through misuse of explosives. Since then decades of research resulted in two large meta-analyses, which showed consistent immune dysregulation in healthy people who are experiencing stress.

In the first meta-analysis by Herbert and Cohen in 1993, they examined 38 studies of stressful events and immune function in healthy adults. They included studies of acute laboratory stressors (e.g. a speech task), short-term naturalistic stressors (e.g. medical examinations), and long-term naturalistic stressors (e.g. divorce, bereavement, caregiving, unemployment). They found consistent stress-related increases in numbers of total white blood cells, as well as decreases in the numbers of helper T cells, suppressor T cells, and cytotoxic T cells, B cells, and Natural killer cells (NK). They also reported stress-related decreases in NK and T cell function, and T cell proliferative responses to phytohaemagglutinin [PHA] and concanavalin A [Con A]. These effects were consistent for short-term and long-term naturalistic stressors, but not laboratory stressors.

In the second meta-analysis by Zorrilla et al. in 2001, they replicated Herbert and Cohen's meta-analysis. Using the same study selection procedures, they analyzed 75 studies of stressors and human immunity. Naturalistic stressors were associated with increases in number of circulating neutrophils, decreases in number and percentages of total T cells and helper T cells, and decreases in percentages of Natural killer cell (NK) cells and cytotoxic T cell lymphocytes. They also replicated Herbert and Cohen's finding of stress-related decreases in NKCC and T cell mitogen proliferation to Phytohaemagglutinin (PHA) and Concanavalin A (Con A).

More recently, there has been increasing interest in the links between interpersonal stressors and immune function. For example, marital conflict, loneliness, caring for a person with a chronic medical condition, and other forms on interpersonal stress dysregulate immune function.

Communication between The Brain and Immune System

- Stimulation of brain sites alters immunity (stressed animals have altered immune systems).
- Damage to brain hemispheres alters immunity (hemispheric lateralization effects).

- Immune cells produce cytokines that act on the CNS.
- Immune cells respond to signals from the CNS.

Communication between Neuroendocrine and Immune System

- Glucocorticoids and catecholamines influence immune cells.
- Endorphins from pituitary & adrenal medulla act on immune system.
- Activity of the immune system is correlated with neurochemical/neuroendocrine activity of brain cells.

Connections between Glucocorticoids and Immune System

- Anti-inflammatory hormones that enhance the organism's response to a stressor.
- Prevent the overreaction of the body's own defense system.
- Regulators of the immune system.
- Affect cell growth, proliferation & differentiation.
- Cause immunosuppression.
- Suppress cell adhesion, antigen presentation, chemotaxis & cytotoxicity.
- Increase apoptosis.

Corticotropin-Releasing Hormone (CRH)

Release of corticotropin-releasing hormone (CRH) from the hypothalamus is influenced by stress.

- CRH is a major regulator of the HPA axis/stress axis.
- CRH Regulates secretion of Adrenocorticotropic hormone (ACTH).
- CRH is widely distributed in the brain and periphery
- CRH also regulates the actions of the Autonomic nervous system ANS and immune system.

Furthermore, stressors that enhance the release of CRH suppress the function of the immune system; conversely, stressors that depress CRH release potentiate immunity.

- Central mediated since peripheral administration of CRH antagonist does not affect immunosuppression.

Pharmaceutical Advances

Glutamate agonists, cytokine inhibitors, vanilloid-receptor agonists, catecholamine modulators, ion-channel blockers, anticonvulsants, GABA agonists (including opioids and cannabinoids), COX inhibitors, acetylcholine modulators, melatonin analogs (such as Ramelton), adenosine receptor antagonists and several miscellaneous drugs (including biologics like Passiflora edulis) are being studied for their psychoneuroimmunological effects.

For example, SSRIs, SNRIs and tricyclic antidepressants acting on serotonin, norepinephrine and dopamine receptors have been shown to be immunomodulatory and anti-inflammatory against pro-inflammatory cytokine processes, specifically on the regulation of IFN-gamma and IL-10, as well as TNF-alpha and IL-6 through a psychoneuroimmunological process. Antidepressants have also been shown to suppress TH1 upregulation.

Tricyclic and dual serotonergic-noradrenergic reuptake inhibition by SNRIs (or SSRI-NRI combinations), have also shown analgesic properties additionally. According to recent evidences antidepressants also seem to exert beneficial effects in experimental autoimmune neuritis in rats by decreasing Interferon-beta (IFN-beta) release or augmenting NK activity in depressed patients.

These studies warrant investigation for antidepressants for use in both psychiatric and non-psychiatric illness and that a psychoneuroimmunological approach may be required for optimal pharmacotherapy in many diseases. Future antidepressants may be made to specifically target the immune system by either blocking the actions of pro-inflammatory cytokines or increasing the production of anti-inflammatory cytokines.

Extrapolating from the observations that positive emotional experiences boost the immune system, Roberts speculates that intensely positive emotional experiences —sometimes brought about during mystical experiences occasioned by psychedelic medicines—may boost the immune system powerfully. Research on salivary IgA supports this hypothesis, but experimental testing has not been done.

Osteoimmunology

Osteoimmunology is a field that emerged about 40 years ago that studies the in-terface between the skeletal system and the immune system, comprising the "osteo-immune system". Osteoimmunology also studies the shared components and mechanisms between the two systems in vertebrates, including ligands, receptors, signaling molecules and transcription factors. Over the past decade, osteoimmunology has been investigated clinically for the treatment of bone metastases, rheumatoid arthritis (RA), osteoporosis, osteopetrosis, and periodontitis. Studies in osteoimmunology reveal relationships between molecular communication among blood cells and structural pathologies in the body.

System Similarities

The RANKL-RANK-OPG (osteoprotegerin) axis is an example of an important signaling system functioning both in bone and immune cell communication. RANKL is expressed on osteoblasts and activated T cells, whereas RANK is expressed on osteoclasts, and dendritic cells (DCs), both of which can be derived from myeloid progenitor cells. Surface RANKL on osteoblasts as well as secreted RANKL provide necessary signals for osteoclast precursors to differentiate into osteoclasts. RANKL expression on activated T cells leads to DC activation through binding to RANK expressed on DCs. OPG, produced by DCs, is a soluble decoy receptor for RANKL that competitively inhibits RANKL binding to RANK.

Crosstalk

The bone marrow cavity is important for the proper development of the immune system, and houses important stem cells for maintenance of the immune system. Within this space, as well as outside of it, cytokines produced by immune cells also have important effects on regulating bone homeostasis. Some important cytokines that are produced by the immune system, including RANKL, M-CSF, TNFa, ILs, and IFNs, affect the differentiation and activity of osteoclasts and bone resorption. Such inflammatory osteoclastogenesis and osteoclast activation can be seen in ex vivo primary cultures of cells from the inflamed synovial fluid of patients with disease flare of the autoimmune disease rheumatoid arthritis.

Clinical Osteoimmunology

Clinical osteoimmunology is a field that studies a treatment or prevention of the bone related diseases caused by disorders of the immune system. Aberrant and/or prolonged activation of immune system leads to derangement of bone modeling and remodeling. Common diseases caused by disorder of osteoimmune system is osteoporosis and bone destruction accompanied by RA characterized by high infiltration of CD4+ T cells in rheumatoid joints, in which two mechanisms are involved: One is an indirect effect on osteoclastogenesis from rheumatoid synovial cells in joints since synovial cells have osteoclast precursors and osteoclast supporting cells, synovial macrophages are highly differentiated into osteoclasts with help of RANKL released from osteoclast supporting cells. The second is an indirect effect on osteoclast differentiation and activity by the secretion of inflammatory cytokines such as IL-1, IL-6, TNFa, in synovium of RA, which increase RANKL signaling and finally bone destruction. A clinical approach to prevent bone related diseases caused by RA is OPG and RANKL treatment in arthritis.

References

- Zitvogel L, Casares N, Péquignot MO, Chaput N, Albert ML, Kroemer G (2004). "Immune response against dying tumor cells". Advances in Immunology. Advances in Immunology. 84: 131–79. doi:10.1016/S0065-2776(04)84004-5. ISBN 978-0-12-022484-5. PMID 15246252.

- Hedger MP, Hales DB (2006). "Immunophysiology of the Male Reproductive Tract". In Neill JD. Knobil and Neill's Physiology of Reproduction. 1 (3rd ed.). Elsevier. pp. 1195–1286. ISBN 978-0-12-515401-7.

- Buckwalter MR, Srivastava PK (2013). "Mechanism of dichotomy between CD8+ responses elicited by apoptotic and necrotic cells". Cancer Immunity. 13: 2. PMC 3559190. PMID 23390373.

- Gamrekelashvili J, Ormandy LA, Heimesaat MM, Kirschning CJ, Manns MP, Korangy F, Greten TF (Oct 2012). "Primary sterile necrotic cells fail to cross-prime CD8(+) T cells". Oncoimmunology. 1 (7): 1017–1026. doi:10.4161/onci.21098. PMC 3494616. PMID 23170250.

- "Robert Ader, Founder of Psychoneuroimmunology, Dies". University of Rochester Medical Center. 2011-12-20. Retrieved 2011-12-20.

- Sumner R.C.; Parton A.; Nowicky A.N.; Kishore U.; Gidron Y. "Hemispheric lateralisation and immune function: A systematic review of human research". Journal of Neuroimmunology. 240 -241: 1–12. doi:10.1016/j.jneuroim.2011.08.017.

Divisions of Immune System

This chapter deals exclusively with the immune system and its categories. The chapter provides detailed information about the innate immune system, adaptive immune system and complement system. The chapter acquaints the reader with the definition and discerning characteristics of each division.

Innate Immune System

INNATE IMMUNE SYSTEM

Innate immune system

The innate immune system, also known as the non-specific immune system or in-born immunity system, is an important subsystem of the overall immune system that comprises the cells and mechanisms that defend the host from infection by other organisms. The cells of the innate system recognize and respond to pathogens in a generic way, but, unlike the adaptive immune system, the system does not confer long-lasting or protective immunity to the host. Innate immune systems provide immediate defense against infection, and are found in all classes of plant and animal life.

The innate immune system is an evolutionarily older defense strategy, and is the dominant immune system found in plants, fungi, insects, and primitive multicellular organisms.

The major functions of the vertebrate innate immune system include:

- Recruiting immune cells to sites of infection, through the production of chemical factors, including specialized chemical mediators, called cytokines

- Activation of the complement cascade to identify bacteria, activate cells, and promote clearance of antibody complexes or dead cells

- Identification and removal of foreign substances present in organs, tissues, blood and lymph, by specialized white blood cells

- Activation of the adaptive immune system through a process known as antigen presentation

- Acting as a physical and chemical barrier to infectious agents.

Anatomical Barriers

Anatomical barrier	Additional defense mechanisms
Skin	Sweat, desquamation, flushing, organic acids
Gastrointestinal tract	Peristalsis, gastric acid, bile acids, digestive enzyme, flushing, thiocyanate, defensins, gut flora
Respiratory airways and lungs	Mucociliary elevator, surfactant, defensins
Nasopharynx	Mucus, saliva, lysozyme
Eyes	Tears

Anatomical barriers include physical, chemical and biological barriers. The epithelial surfaces form a physical barrier that is impermeable to most infectious agents, acting as the first line of defense against invading organisms. Desquamation (shedding) of skin epithelium also helps remove bacteria and other infectious agents that have adhered to the epithelial surfaces. Lack of blood vessels and inability of the epidermis to retain moisture, presence of sebaceous glands in the dermis provides an environment unsuitable for the survival of microbes. In the gastrointestinal and respiratory tract, movement due to peristalsis or cilia, respectively, helps remove infectious agents. Also, mucus traps infectious agents. The gut flora can prevent the colonization of pathogenic bacteria by secreting toxic substances or by competing with pathogenic bacteria for nutrients or attachment to cell surfaces. The flushing action of tears and saliva helps prevent infection of the eyes and mouth.

Inflammation

Inflammation is one of the first responses of the immune system to infection or irritation. Inflammation is stimulated by chemical factors released by injured cells and serves to establish a physical barrier against the spread of infection, and to promote healing of any damaged tissue following the clearance of pathogens.

The process of acute inflammation is initiated by cells already present in all tissues, mainly resident macrophages, dendritic cells, histiocytes, Kupffer cells, and mastocytes. These cells present receptors contained on the surface or within the cell, named *pattern recognition receptors* (PRRs), which recognize molecules that are broadly shared by pathogens but distinguishable from host molecules, collectively referred to as pathogen-associated molecular patterns (PAMPs). At the onset of an infection, burn, or other injuries, these cells undergo activation (one of their PRRs recognizes a PAMP) and release inflammatory mediators responsible for the clinical signs of inflammation.

Chemical factors produced during inflammation (histamine, bradykinin, serotonin, leukotrienes, and prostaglandins) sensitize pain receptors, cause local vasodilation of the blood vessels, and attract phagocytes, especially neutrophils. Neutrophils then trigger other parts of the immune system by releasing factors that summon additional leukocytes and lymphocytes. Cytokines produced by macrophages and other cells of the innate immune system mediate the inflammatory response. These cytokines include TNF, HMGB1, and IL-1.

The inflammatory response is characterized by the following symptoms:

- redness of the skin, due to locally increased blood circulation;

- heat, either increased local temperature, such as a warm feeling around a localized infection, or a systemic fever;

- swelling of affected tissues, such as the upper throat during the common cold or joints affected by rheumatoid arthritis;

- increased production of mucus, which can cause symptoms like a runny nose or a productive cough;

- pain, either local pain, such as painful joints or a sore throat, or affecting the whole body, such as body aches; and

- possible dysfunction of the organs or tissues involved.

Complement System

The complement system is a biochemical cascade of the immune system that helps, or "complements", the ability of antibodies to clear pathogens or mark them for destruction by other cells. The cascade is composed of many plasma proteins, synthesized in the liver, primarily by hepatocytes. The proteins work together to:

- trigger the recruitment of inflammatory cells

- "tag" pathogens for destruction by other cells by *opsonizing*, or coating, the surface of the pathogen

- form holes in the plasma membrane of the pathogen, resulting in cytolysis of the pathogen cell, causing the death of the pathogen

- rid the body of neutralised antigen-antibody complexes.

Elements of the complement cascade can be found in many non-mammalian species including plants, birds, fish, and some species of invertebrates.

Cells of The Innate Immune Response

All white blood cells (WBCs) are known as leukocytes. Leukocytes differ from other cells of the body in that they are not tightly associated with a particular organ or tissue; thus, their function is similar to that of independent, single-cell organisms. Leukocytes are able to move freely and interact with and capture cellular debris, foreign particles, and invading microorganisms. Unlike many other cells in the body, most innate immune leukocytes cannot divide or reproduce on their own, but are the products of multipotent hematopoietic stem cells present in the bone marrow.

The innate leukocytes include: Natural killer cells, mast cells, eosinophils, basophils; and the phagocytic cells include macrophages, neutrophils, and dendritic cells, and function within the immune system by identifying and eliminating pathogens that might cause infection.

Mast Cells

Mast cells are a type of innate immune cell that reside in connective tissue and in the mucous membranes. They are intimately associated with wound healing and defense against pathogens, but are also often associated with allergy and anaphylaxis. When activated, mast cells rapidly release characteristic granules, rich in histamine and heparin, along with various hormonal mediators and chemokines, or chemotactic cytokines into the environment. Histamine dilates blood vessels, causing the characteristic signs of inflammation, and recruits neutrophils and macrophages.

Phagocytes

The word 'phagocyte' literally means 'eating cell'. These are immune cells that engulf, or 'phagocytose', pathogens or particles. To engulf a particle or pathogen, a phagocyte extends portions of its plasma membrane, wrapping the membrane around the particle until it is enveloped (i.e., the particle is now inside the cell). Once inside the cell, the invading pathogen is contained inside an endosome, which merges with a lysosome. The lysosome contains enzymes and acids that kill and digest the particle or organism. In general, phagocytes patrol the body searching for pathogens, but are also able to react to a group of highly specialized molecular signals produced by other cells, called cytokines. The phagocytic cells of the immune system include macrophages, neutrophils, and dendritic cells.

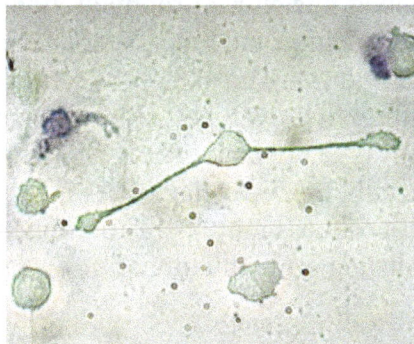

A macrophage

Phagocytosis of the hosts' own cells is common as part of regular tissue development and maintenance. When host cells die, either by programmed cell death (also called apoptosis) or by cell injury due to a bacterial or viral infection, phagocytic cells are responsible for their removal from the affected site. By helping to remove dead cells preceding growth and development of new healthy cells, phagocytosis is an important part of the healing process following tissue injury.

Macrophages

Macrophages, from the Greek, meaning "large eaters," are large phagocytic leukocytes, which are able to move outside of the vascular system by migrating across the walls of capillary vessels and entering the areas between cells in pursuit of invading pathogens. In tissues, organ-specific macrophages are differentiated from phagocytic cells present in the blood called monocytes. Macrophages are the most efficient phagocytes and can phagocytose substantial numbers of bacteria or other cells or microbes. The binding of bacterial molecules to receptors on the surface of a macrophage triggers it to engulf and destroy the bacteria through the generation of a "respiratory burst", causing the release of reactive oxygen species. Pathogens also stimulate the macrophage to produce chemokines, which summon other cells to the site of infection.

Neutrophils

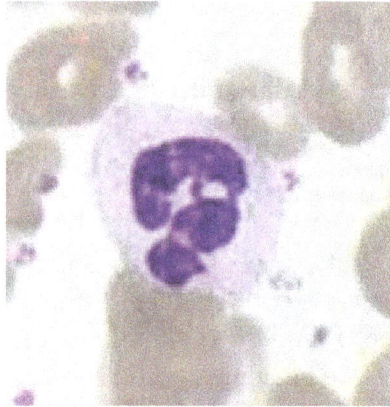

A neutrophil

Neutrophils, along with two other cell types (eosinophils and basophils), are known as granulocytes due to the presence of granules in their cytoplasm, or as polymorphonuclear cells (PMNs) due to their distinctive lobed nuclei. Neutrophil granules contain a variety of toxic substances that kill or inhibit growth of bacteria and fungi. Similar to macrophages, neutrophils attack pathogens by activating a respiratory burst. The main products of the neutrophil respiratory burst are strong oxidizing agents including hydrogen peroxide, free oxygen radicals and hypochlorite. Neutrophils are the most abundant type of phagocyte, normally representing 50-60% of the total circulating leukocytes, and are usually the first cells to arrive at the site of an infection. The bone marrow of a normal healthy adult produces more than 100 billion neutrophils per day, and more than 10 times that many per day during acute inflammation.

Dendritic Cells

Dendritic cells (DCs) are phagocytic cells present in tissues that are in contact with the external environment, mainly the skin (where they are often called Langerhans cells), and the inner mucosal lining of the nose, lungs, stomach, and intestines. They are named for their resemblance to neuronal dendrites, but dendritic cells are not connected to the nervous system. Dendritic cells are very important in the process of antigen presentation, and serve as a link between the innate and adaptive immune systems.

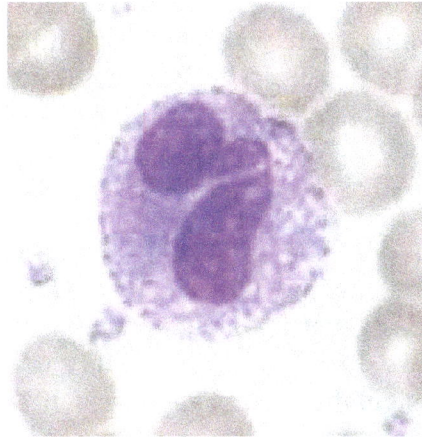

An eosinophil

Basophils and Eosinophils

Basophils and eosinophils are cells related to the neutrophil. When activated by a pathogen encounter, histamine-releasing basophils are important in the defense against parasites and play a role in allergic reactions, such as asthma. Upon activation, eosinophils secrete a range of highly toxic proteins and free radicals that are highly effective in killing parasites, but may also damage tissue during an allergic reaction. Activation and release of toxins by eosinophils are, therefore, tightly regulated to prevent any inappropriate tissue destruction.

Natural Killer Cells

Natural killer cells (NK cells) are a component of the innate immune system that does not directly attack invading microbes. Rather, NK cells destroy compromised host cells, such as tumor cells or virus-infected cells, recognizing such cells by a condition known as "missing self." This term describes cells with abnormally low levels of a cell-surface marker called MHC I (major histocompatibility complex) - a situation that can arise in viral infections of host cells. They were named "natural killer" because of the initial notion that they do not require activation in order to kill cells that are "missing self." For many years, it was unclear how NK cell recognize tumor cells and infected cells. It is now known that the MHC makeup on the surface of those cells is altered and the NK cells become activated through recognition of "missing self". Normal body cells are not recognized and attacked by NK cells because they express intact self MHC antigens. Those MHC antigens are recognized by killer cell immunoglobulin receptors (KIR) that, in essence, put the brakes on NK cells. The NK-92 cell line does not express KIR and is developed for tumor therapy.

γδ T cells

Like other 'unconventional' T cell subsets bearing invariant T cell receptors (TCRs), such as CD1d-restricted Natural Killer T cells, γδ T cells exhibit characteristics that place them at the border between innate and adaptive immunity. On one hand, γδ T cells may be considered a component of adaptive immunity in that they rearrange TCR genes to produce junctional diversity and develop a memory phenotype. However, the various subsets may also be considered part of the innate immune system where a restricted TCR or NK receptors may be used as a pattern recognition receptor. For example, according to this paradigm, large numbers of Vγ9/Vδ2 T cells respond within hours to common molecules produced by microbes, and highly restricted intraepithelial Vδ1 T cells will respond to stressed epithelial cells.

Other Vertebrate Mechanisms

The coagulation system overlaps with the immune system. Some products of the coagulation system can contribute to the non-specific defenses by their ability to increase vascular permeability and act as chemotactic agents for phagocytic cells. In addition, some of the products of the coagulation system are directly antimicrobial. For example, beta-lysine, a protein produced by platelets during coagulation, can cause lysis of many Gram-positive bacteria by acting as a cationic detergent. Many acute-phase proteins of inflammation are involved in the coagulation system.

Also increased levels of lactoferrin and transferrin inhibit bacterial growth by binding iron, an essential nutrient for bacteria.

Neural Regulation of Innate Immunity

The innate immune response to infectious and sterile injury is modulated by neural circuits that control cytokine production period. The Inflammatory Reflex is a prototypical neural circuit that controls cytokine production in spleen. Action potentials transmitted via the vagus nerve to spleen mediate the release of acetylcholine, the neurotransmitter that inhibits cytokine release by interacting with alpha7 nicotinic acetylcholine receptors (CHRNA7) expressed on cytokine-producing cells. The motor arc of the inflammatory reflex is termed the cholinergic anti-inflammatory pathway.

Pathogen-Specificity

The parts of the innate immune system have different specificity for different pathogens.

Pathogen	Main examples	Phagocytosis	complement	NK cells
Intracellular and cytoplasmic virus	• influenza • mumps • measles • rhinovirus	yes	no	yes
Intracellular bacteria	• Listeria monocytogenes • Legionella • Mycobacterium • Rickettsia	yes (specifically neutrophils, no for rickettsia)	no	yes (no for rickettsia)

Extracellular bacteria	• Staphylococcus • Streptococcus • Neisseria • Salmonella typhi	yes	yes	no
Intracellular protozoa	• Plasmodium malariae • Leishmania donovani	no	no	no
Extracellular protozoa	• Entamoeba histolytica • Giardia lamblia	yes	yes	no
Extracellular fungi	• Candida • Histoplasma • Cryptococcus	no	yes	yes

Innate Immune Evasion

Cells of the innate immune system, in effect, prevent free growth of bacteria within the body; however, many pathogens have evolved mechanisms allowing them to evade the innate immune system.

Evasion strategies that circumvent the innate immune system include intracellular replication, such as in *Mycobacterium tuberculosis*, or a protective capsule that prevents lysis by complement and by phagocytes, as in *salmonella*. *Bacteroides* species are normally mutualistic bacteria, making up a substantial portion of the mammalian gastrointestinal flora. Some species (*B. fragilis*, for example) are opportunistic pathogens, causing infections of the peritoneal cavity. These species evade the immune system through inhibition of phagocytosis by affecting the receptors that phagocytes use to engulf bacteria or by mimicking host cells so that the immune system does not recognize them as foreign. *Staphylococcus aureus* inhibits the ability of the phagocyte to respond to chemokine signals. Other organisms such as *M. tuberculosis*, *Streptococcus pyogenes*, and *Bacillus anthracis* utilize mechanisms that directly kill the phagocyte.

Bacteria and fungi may also form complex biofilms, providing protection from the cells and proteins of the immune system; recent studies indicate that such biofilms are present in many successful infections, including the chronic *Pseudomonas aeruginosa* and *Burkholderia cenocepacia* infections characteristic of cystic fibrosis.

Evasion of The Innate Immune System by Virus.

Type I interferons (IFN), secreted mainly by dendritic cells, play the central role in antiviral host defense and creation of an effective antiviral state in a cell. Viral components are recognized by different receptors: Toll-like receptors are located in the endosomal membrane and recognize double-stranded RNA (dsRNA), MDA5 and RIG-I receptors are located in the cytoplasm and recognize long dsRNA and phosphate-containing dsRNA respectively. The viral recognition by MDA5 and RIG-I receptors in the cytoplasm induces a conformational change between the caspase-recruitment domain (CARD) and the CARD-containing adaptor MAVS.

In parallel, the viral recognition by toll-like receptors in the endocytic compartments induces the activation of the adaptor protein TRIF. These two pathways converge in the recruitment and activation of the IKKε/TBK-1 complex, inducing phosphorylation and homo- and hetero-dimerization of transcription factors IRF3 and IRF7. These molecules are translocated in the nucleus, where they induce IFN production with the presence of C-Jun (a particular transcription factor) and activating transcription factor 2. IFN then binds to the IFN receptors, inducing expression of hundreds of interferon-stimulated genes. This leads to production of proteins with antiviral properties, such as protein kinase R, which inhibits viral protein synthesis, or the 2′,5′-oligoadenylate synthetase family, which degrades viral RNA. These molecules establish an antiviral state in the cell.

Some viruses are able to evade this immune system by producing molecules that interfere with the IFN production pathway. For example, the Influenza A virus produces NS1 protein, which can bind to single-stranded and double-stranded RNA, thus inhibiting type I IFN production. Influenza A virus also blocks protein kinase R activation and the establishment of the antiviral state. The dengue virus also inhibits type I IFN production by blocking IRF-3 phosophorylation using NS2B3 protease complex.

Innate Immunity in Other Species

Host Defense in Prokaryotes

Bacteria (and perhaps other prokaryotic organisms), utilize a unique defense mechanism, called the restriction modification system to protect themselves from pathogens, such as bacteriophages. In this system, bacteria produce enzymes, called restriction endonucleases, that attack and destroy specific regions of the viral DNA of invading bacteriophages. Methylation of the host's own DNA marks it as "self" and prevents it from being attacked by endonucleases. Restriction endonucleases and the restriction modification system exist exclusively in prokaryotes.

Host Defense in Invertebrates

Invertebrates do not possess lymphocytes or an antibody-based humoral immune system, and it is likely that a multicomponent, adaptive immune system arose with the first vertebrates. Nevertheless, invertebrates possess mechanisms that appear to be precursors of these aspects of vertebrate immunity. *Pattern recognition receptors* are proteins used by nearly all organisms to identify molecules associated with microbial pathogens. *Toll-like receptors* are a major class of pattern recognition receptor, that exists in all coelomates (animals with a body-cavity), including humans. The complement system, as discussed above, is a biochemical cascade of the immune system that helps clear pathogens from an organism, and exists in most forms of life. Some invertebrates, including various insects, crabs, and worms utilize a modified form of the complement response known as the prophenoloxidase (proPO) system.

Antimicrobial peptides are an evolutionarily conserved component of the innate immune response found among all classes of life and represent the main form of invertebrate systemic immunity. Several species of insect produce antimicrobial peptides known as *defensins* and *cecropins*.

Proteolytic Cascades

In invertebrates, pattern recognition proteins (PRPs) trigger proteolytic cascades that degrade proteins and control many of the mechanisms of the innate immune system of invertebrates—including hemolymph coagulation and melanization. Proteolytic cascades are important components of the invertebrate immune system because they are turned on more rapidly than other innate immune reactions because they do not rely on gene changes. Proteolytic cascades have been found to function the same in both vertebrate and invertebrates, even though different proteins are used throughout the cascades.

Clotting Mechanisms

In the hemolymph, which makes up the fluid in the circulatory system of arthropods, a gel-like fluid surrounds pathogen invaders, similar to the way blood does in other animals. There are various different proteins and mechanisms that are involved in invertebrate clotting. In crustaceans, transglutaminase from blood cells and mobile plasma proteins make up the clotting system, where the transglutaminase polymerizes 210 kDa subunits of a plasma-clotting protein. On the other hand, in the horseshoe crab species clotting system, components of proteolytic cascades are stored as inactive forms in granules of hemocytes, which are released when foreign molecules, like lipopolysaccharides enter.

Host Defense in Plants

Members of every class of pathogen that infect humans also infect plants. Although the exact pathogenic species vary with the infected species, bacteria, fungi, viruses, nematodes, and insects can all cause plant disease. As with animals, plants attacked by insects or other pathogens use a set of complex metabolic responses that lead to the formation of defensive chemical compounds that fight infection or make the plant less attractive to insects and other herbivores.

Like invertebrates, plants neither generate antibody or T-cell responses nor possess mobile cells that detect and attack pathogens. In addition, in case of infection, parts of some plants are treated as disposable and replaceable, in ways that very few animals are able to do. Walling off or discarding a part of a plant helps stop spread of an infection.

Most plant immune responses involve systemic chemical signals sent throughout a plant. Plants use pattern-recognition receptors to recognize conserved microbial signatures. This recognition triggers an immune response. The first plant receptors of conserved microbial signatures were identified in rice (XA21, 1995) and in Arabidopsis (FLS2, 2000). Plants also carry immune receptors that recognize highly variable pathogen effectors. These include the NBS-LRR class of proteins. When a part of a plant becomes infected with a microbial or viral pathogen, in case of an incompatible interaction triggered by specific elicitors, the plant produces a localized hypersensitive response (HR), in which cells at the site of infection undergo rapid programmed cell death to prevent the spread of the disease to other parts of the plant. HR has some similarities to animal pyroptosis, such as a requirement of caspase-1-like proteolytic activity of VPEγ, a cysteine protease that regulates cell disassembly during cell death.

"Resistance" (R) proteins, encoded by R genes, are widely present in plants and detect pathogens. These proteins contain domains similar to the NOD Like Receptors and Toll-like receptors utilized in animal innate immunity. Systemic acquired resistance (SAR) is a type of defensive response that renders the entire plant resistant to a broad spectrum of infectious agents. SAR involves the production of chemical messengers, such as salicylic acid or jasmonic acid. Some of these travel through the plant and signal other cells to produce defensive compounds to protect uninfected parts, e.g., leaves. Salicylic acid itself, although indispensable for expression of SAR, is not the translocated signal responsible for the systemic response. Recent evidence indicates a role for jasmonates in transmission of the signal to distal portions of the plant. RNA silencing mechanisms are also important in the plant systemic response, as they can block virus replication. The *jasmonic acid response*, is stimulated in leaves damaged by insects, and involves the production of methyl jasmonate.

Adaptive Immune System

The adaptive immune system, also known as the acquired immune system or, more rarely, as the specific immune system, is a subsystem of the overall immune system that is composed of highly specialized, systemic cells and processes that eliminate or prevent pathogen growth. The adaptive immune system is one of the two main immunity strategies found in vertebrates (the other being the innate immune system). Adaptive immunity creates immunological memory after an initial response to a specific pathogen, and leads to an enhanced response to subsequent encounters with that pathogen. This process of acquired immunity is the basis of vaccination. Like the innate system, the adaptive system includes both humoral immunity components and cell-mediated immunity components.

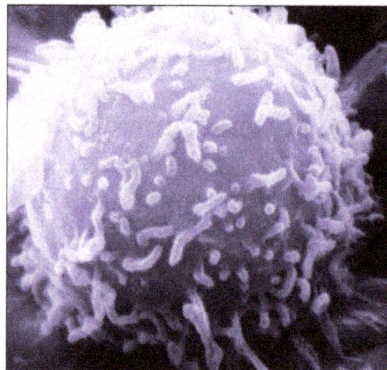

A scanning electron microscope (SEM) image of a single human lymphocyte

Unlike the innate immune system, the adaptive immune system is highly specific to a particular pathogen. Adaptive immunity can also provide long-lasting protection: for example; someone who recovers from measles is now protected against measles for their lifetime but in other cases it does not provide lifetime protection: for example; chickenpox. The adaptive system response destroys invading pathogens and any toxic molecules they produce. Sometimes the adaptive system is unable to distinguish harmful from harmless foreign molecules; the effects of this may be hayfever, asthma or any other allergies. Antigens are any substances that elicit the adaptive immune response. The cells that carry out the adaptive immune response are white blood cells known as lymphocytes. Two main broad classes—antibody responses and cell mediated immune response— are also carried by two different lymphocytes (B cells and T cells). In antibody responses, B cells are activated to secrete antibodies, which are proteins also known as immunoglobulins. Antibodies travel through the bloodstream and bind to the foreign antigen causing it to inactivate, which does not allow the antigen to bind to the host.

In acquired immunity, pathogen-specific receptors are "acquired" during the lifetime of the organism (whereas in innate immunity pathogen-specific receptors are already encoded in the germline). The acquired response is called "adaptive" because it prepares the body's immune system for future challenges (though it can actually also be maladaptive when it results in autoimmunity).

The system is highly adaptable because of somatic hypermutation (a process of accelerated somatic mutations), and V(D)J recombination (an irreversible genetic recombination of antigen receptor gene segments). This mechanism allows a small number of genes to generate a vast number of different antigen receptors, which are then uniquely expressed on each individual lymphocyte. Since the gene rearrangement leads to an irreversible change in the DNA of each cell, all progeny (offspring) of that cell inherit genes that encode the same receptor specificity, including the memory B cells and memory T cells that are the keys to long-lived specific immunity.

A theoretical framework explaining the workings of the acquired immune system is provided by immune network theory. This theory, which builds on established concepts of clonal selection, is being applied in the search for an HIV vaccine.

Functions

Acquired immunity is triggered in vertebrates when a pathogen evades the innate immune system and (1) generates a threshold level of antigen and (2) generates "stranger" or "danger" signals activating dendritic cells.

The major functions of the acquired immune system include:

- Recognition of specific "non-self" antigens in the presence of "self", during the process of antigen presentation.

- Generation of responses that are tailored to maximally eliminate specific pathogens or pathogen-infected cells.

- Development of immunological memory, in which pathogens are "remembered" through memory B cells and memory T cells.

Lymphocytes

The cells of the acquired immune system are T and B lymphocytes; lymphocytes are a subset of leukocyte. B cells and T cells are the major types of lymphocytes. The human body has about 2 trillion lymphocytes, constituting 20–40% of white blood cells (WBCs); their total mass is about the same as the brain or liver. The peripheral blood contains 2% of circulating lymphocytes; the rest move within the tissues and lymphatic system.

B cells and T cells are derived from the same multipotent hematopoietic stem cells, and are morphologically indistinguishable from one another until after they are activated. B cells play a large role in the humoral immune response, whereas T cells are intimately involved in cell-mediated immune responses. In all vertebrates except Agnatha, B cells and T cells are produced by stem cells in the bone marrow.

T progenitors migrate from the bone marrow to the thymus where they are called thymocytes and where they develop into T cells. In humans, approximately 1–2% of the lymphocyte pool recirculates each hour to optimize the opportunities for antigen-specific lymphocytes to find their specific antigen within the secondary lymphoid tissues.

In an adult animal, the peripheral lymphoid organs contain a mixture of B and T cells in at least three stages of differentiation:

- naive B and naive T cells (cells that have not matured), left the bone marrow or thymus, have entered the lymphatic system, but have yet to encounter their cognate antigen,

- effector cells that have been activated by their cognate antigen, and are actively involved in eliminating a pathogen.

- memory cells – the long-lived survivors of past infections.

Antigen Presentation

Acquired immunity relies on the capacity of immune cells to distinguish between the body's own cells and unwanted invaders. The host's cells express "self" antigens. These antigens are different from those on the surface of bacteria or on the surface of virus-infected host cells ("non-self" or "foreign" antigens). The acquired immune response is triggered by recognizing foreign antigen in the cellular context of a danger-activated dendritic cell.

With the exception of non-nucleated cells (including erythrocytes), all cells are capable of presenting antigen through the function of major histocompatibility complex (MHC) molecules. Some cells are specially equipped to present antigen, and to prime naive T cells. Dendritic cells and B-cells (and to a lesser extent macrophages) are equipped with special "co-stimulatory" ligands recognized by co-stimulatory receptors on T cells, and are termed professional antigen-presenting cells (APC).

Several T cells subgroups can be activated by professional APCs, and each type of T cell is specially equipped to deal with each unique toxin or bacterial and viral pathogen. The type of T cell activated, and the type of response generated, depends, in part, on the context in which the APC first encountered the antigen.

Exogenous Antigens

Dendritic cells engulf exogenous pathogens, such as bacteria, parasites or toxins in the tissues and then migrate, via chemotactic signals, to the T cell-enriched lymph nodes. During migration, dendritic cells undergo a process of maturation in which they lose most of their ability to engulf other pathogens and develop an ability to communicate with T-cells. The dendritic cell uses enzymes to chop the pathogen into smaller pieces, called antigens. In the lymph node, the dendritic cell displays these "non-self" antigens on its surface by coupling them to a "self"-receptor called the major histocompatibility complex, or MHC (also known in humans as human leukocyte antigen (HLA)). This MHC:antigen complex is recognized by T-cells passing through the lymph node. Exogenous antigens are usually displayed on MHC class II molecules, which activate CD4+T helper cells.

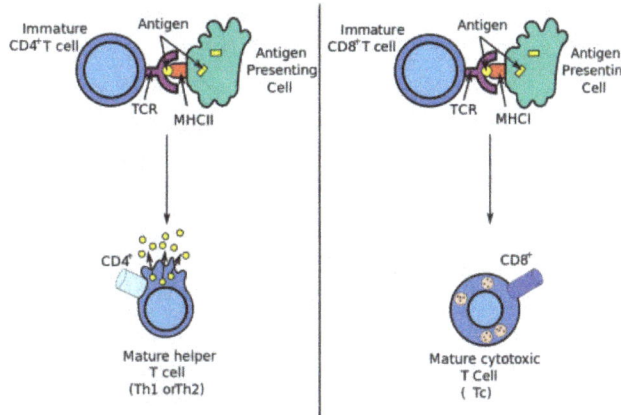

Antigen presentation stimulates T cells to become either "cytotoxic" CD8+ cells or "helper" CD4+ cells.

Endogenous Antigens

Endogenous antigens are produced by intracellular bacteria and viruses replicating within a host cell.The host cell uses enzymes to digest virally associated proteins, and displays these pieces on its surface to T-cells by coupling them to MHC. Endogenous antigens are typically displayed on MHC class I molecules, and activate CD8+ cytotoxic T-cells. With the exception of non-nucleated cells (including erythrocytes), MHC class I is expressed by all host cells.

T Lymphocytes

CD8+ T Lymphocytes and Cytotoxicity

Cytotoxic T cells (also known as TC, killer T cell, or cytotoxic T-lymphocyte (CTL)) are a sub-group of T cells that induce the death of cells that are infected with viruses (and other pathogens), or are otherwise damaged or dysfunctional.

Naive cytotoxic T cells are activated when their T-cell receptor (TCR) strongly interacts with a peptide-bound MHC class I molecule. This affinity depends on the type and orientation of the antigen/MHC complex, and is what keeps the CTL and infected cell bound together. Once activated, the CTL undergoes a process called clonal selection, in which it gains functions and divides rapidly to produce an army of "armed" effector cells. Activated CTL then travels throughout the body searching for cells that bear that unique MHC Class I + peptide.

When exposed to these infected or dysfunctional somatic cells, effector CTL release perforin and granulysin: cytotoxins that form pores in the target cell's plasma membrane, allowing ions and water to flow into the infected cell, and causing it to burst or lyse. CTL release granzyme, a serine protease that enters cells via pores to induce apoptosis (cell death). To limit extensive tissue damage during an infection, CTL activation is tightly controlled and in general requires a very strong MHC/antigen activation signal, or additional activation signals provided by "helper" T-cells (see below).

On resolution of the infection, most effector cells die and phagocytes clear them away—but a few of these cells remain as memory cells. On a later encounter with the same antigen, these memory cells quickly differentiate into effector cells, dramatically shortening the time required to mount an effective response.

Helper T-Cells

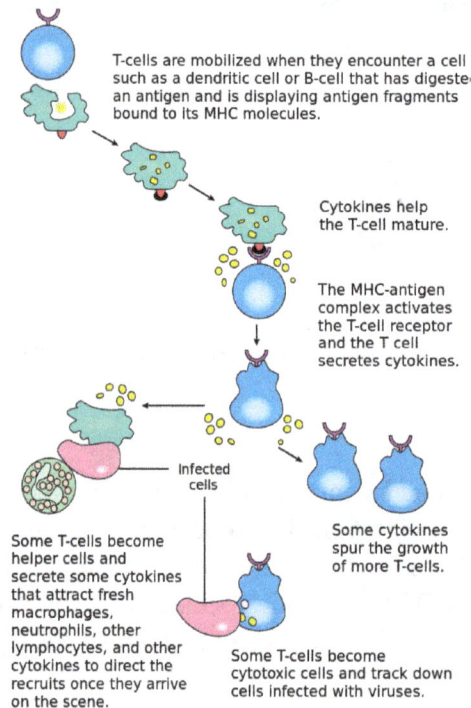

T-cells are mobilized when they encounter a cell such as a dendritic cell or B-cell that has digested an antigen and is displaying antigen fragments bound to its MHC molecules.

Cytokines help the T-cell mature.

The MHC-antigen complex activates the T-cell receptor and the T cell secretes cytokines.

Infected cells

Some cytokines spur the growth of more T-cells.

Some T-cells become helper cells and secrete some cytokines that attract fresh macrophages, neutrophils, other lymphocytes, and other cytokines to direct the recruits once they arrive on the scene.

Some T-cells become cytotoxic cells and track down cells infected with viruses.

The T lymphocyte activation pathway. T cells contribute to immune defenses in two major ways: some direct and regulate immune responses; others directly attack infected or cancerous cells.

CD4+ lymphocytes, also called "helper" or "regulatory" T cells, are immune response mediators, and play an important role in establishing and maximizing the capabilities of the acquired immune response. These cells have no cytotoxic or phagocytic activity; and cannot kill infected cells or clear pathogens, but, in essence "manage" the immune response, by directing other cells to perform these tasks.

Helper T cells express T cell receptors (TCR) that recognize antigen bound to Class II MHC molecules. The activation of a naive helper T-cell causes it to release cytokines, which influences the activity of many cell types, including the APC (Antigen-Presenting Cell) that activated it. Helper T-cells require a much milder activation stimulus than cytotoxic T cells. Helper T cells can provide extra signals that "help" activate cytotoxic cells.

Th1 and Th2: Helper T Cell Responses

Classically, two types of effector CD4+ T helper cell responses can be induced by a professional APC, designated Th1 and Th2, each designed to eliminate different types of pathogens. The factors that dictate whether an infection triggers a Th1 or Th2 type response are not fully understood, but the response generated does play an important role in the clearance of different pathogens.

The Th1 response is characterized by the production of Interferon-gamma, which activates the bactericidal activities of macrophages, and induces B cells to make opsonizing (coating) and complement-fixing antibodies, and leads to *cell-mediated immunity*. In general, Th1 responses are more effective against intracellular pathogens (viruses and bacteria that are inside host cells).

The Th2 response is characterized by the release of Interleukin 5, which induces eosinophils in the clearance of parasites. Th2 also produce Interleukin 4, which facilitates B cell isotype switching. In general, Th2 responses are more effective against extracellular bacteria, parasites including helminths and toxins. Like cytotoxic T cells, most of the CD4+ helper cells die on resolution of infection, with a few remaining as CD4+ memory cells.

Increasingly, there is strong evidence from mouse and human-based scientific studies of a broader diversity in CD4+ effector T helper cell subsets. Regulatory T (Treg) cells, have been identified as important negative regulators of adaptive immunity as they limit and suppresses the immune system to control aberrant immune responses to self-antigens; an important mechanism in controlling the development of autoimmune diseases. Follicular helper T (Tfh) cells are another distinct population of effector CD4+ T cells that develop from naive T cells post-antigen activation. Tfh cells are specialized in helping B cell humoral immunity as they are uniquely capable of migrating to follicular B cells in secondary lymphoid organs and provide them positive paracrine signals to enable the generation and recall production of high-quality affinty-matured antibodies. Similar to Tregs, Tfh cells also play a role in immunological tolerance as an abnormal expansion of Tfh cell numbers can lead to unrestricted autoreactive antibody production causing severe systemic autoimmune disorders.

The relevance of CD4+ T helper cells is highlighted during an HIV infection. HIV is able to subvert the immune system by specifically attacking the CD4+ T cells, precisely the cells that could drive the clearance of the virus, but also the cells that drive immunity against all other pathogens encountered during an organism's lifetime.

Gamma Delta T Cells

Gamma delta T cells (γδ T cells) possess an alternative T cell receptor (TCR) as opposed to CD4+ and CD8+ αβ T cells and share characteristics of helper T cells, cytotoxic T cells and natural killer cells. Like other 'unconventional' T cell subsets bearing invariant TCRs, such as CD1d-restricted natural killer T cells, γδ T cells exhibit characteristics that place them at the border between innate and acquired immunity. On one hand, γδ T cells may be considered a component of acquired immunity in that they rearrange TCR genes via V(D)J recombination, which also produces junctional diversity, and develop a memory phenotype. On the other hand, however, the various subsets may also be considered part of the innate immune system where a restricted TCR or NK receptors may be used as a pattern recognition receptor. For example, according to this paradigm, large numbers of Vγ9/Vδ2 T cells respond within hours to common molecules produced by microbes, and highly restricted intraepithelial Vδ1 T cells respond to stressed epithelial cells.

B Lymphocytes and Antibody Production

B Cells are the major cells involved in the creation of antibodies that circulate in blood plasma and lymph, known as humoral immunity. Antibodies (also known as immunoglobulin, Ig), are large Y-shaped proteins used by the immune system to identify and neutralize foreign objects. In mammals, there are five types of antibody: IgA, IgD, IgE, IgG, and IgM, differing in biological properties; each has evolved to handle different kinds of antigens. Upon activation, B cells produce antibodies, each of which recognizing a unique antigen, and neutralizing specific pathogens.

Like the T cell, B cells express a unique B cell receptor (BCR), in this case, a membrane-bound antibody molecule. All the BCR of any one clone of B cells recognizes and binds to only one particular antigen. A critical difference between B cells and T cells is how each cell "sees" an antigen. T cells recognize their cognate antigen in a processed form – as a peptide in the context of an MHC molecule, whereas B cells recognize antigens in their native form. Once a B cell encounters its cognate (or specific) antigen (and receives additional signals from a helper T cell (predominately Th2 type)), it further differentiates into an effector cell, known as a plasma cell.

Plasma cells are short-lived cells (2–3 days) that secrete antibodies. These antibodies bind to antigens, making them easier targets for phagocytes, and trigger the complement cascade. About 10% of plasma cells survive to become long-lived antigen-specific memory B cells. Already primed to produce specific antibodies, these cells can be called upon to respond quickly if the same pathogen re-infects the host, while the host experiences few, if any, symptoms.

Alternative Acquired Immune System

Although the classical molecules of the acquired immune system (e.g., antibodies and T cell receptors) exist only in jawed vertebrates, a distinct lymphocyte-derived molecule has been discovered in primitive jawless vertebrates, such as the lamprey and hagfish. These animals possess a large array of molecules called variable lymphocyte receptors (VLRs for short) that, like the antigen receptors of jawed vertebrates, are produced from only a small number (one or two) of genes. These molecules are believed to bind pathogenic antigens in a similar way to antibodies, and with the same degree of specificity.

Immunological Memory

When B cells and T cells are activated some become memory B cells and some memory T cells. Throughout the lifetime of an animal these memory cells form a database of effective B and T lymphocytes. Upon interaction with a previously encountered antigen, the appropriate memory cells are selected and activated. In this manner, the second and subsequent exposures to an antigen produce a stronger and faster immune response. This is "adaptive" because the body's immune system prepares itself for future challenges, but is "maladaptive" of course if the receptors are autoimmune. Immunological memory can be in the form of either *passive* short-term memory or *active* long-term memory.

Passive Memory

Passive memory is usually short-term, lasting between a few days and several months. Newborn infants have had no prior exposure to microbes and are particularly vulnerable to infection. Several layers of passive protection are provided by the mother. *In utero*, maternal IgG is transported directly across the placenta, so that, at birth, human babies have high levels of antibodies, with the same range of antigen specificities as their mother. Breast milk contains antibodies (mainly IgA) that are transferred to the gut of the infant, protecting against bacterial infections, until the newborn can synthesize its own antibodies.

This is passive immunity because the fetus does not actually make any memory cells or antibodies: It only borrows them. Short-term passive immunity can also be transferred artificially from one individual to another via antibody-rich serum.

Active Memory

In general, active immunity is long-term and can be acquired by infection followed by B cells and T cells activation, or artificially acquired by vaccines, in a process called immunization.

Immunization

Historically, infectious disease has been the leading cause of death in the human population. Over the last century, two important factors have been developed to combat their spread: sanitation and immunization. Immunization (commonly referred to as vaccination) is the deliberate induction of an immune response, and represents the single most effective manipulation of the immune system that scientists have developed. Immunizations are successful because they utilize the immune system's natural specificity as well as its inducibility.

The principle behind immunization is to introduce an antigen, derived from a disease-causing organism, that stimulates the immune system to develop protective immunity against that organism, but that does not itself cause the pathogenic effects of that organism. An antigen (short for *anti*body *gen*erator), is defined as any substance that binds to a specific antibody and elicits an adaptive immune response.

Most viral vaccines are based on live attenuated viruses, whereas many bacterial vaccines are based on acellular components of microorganisms, including harmless toxin components. Many antigens derived from acellular vaccines do not strongly induce an acquired response, and most bacterial vaccines require the addition of *adjuvants* that activate the antigen-presenting cells of the innate immune system to enhance immunogenicity.

Immunological Diversity

Most large molecules, including virtually all proteins and many polysaccharides, can serve as antigens. The parts of an antigen that interact with an antibody molecule or a lymphocyte receptor, are called epitopes, or antigenic determinants. Most antigens contain a variety of epitopes and can stimulate the production of antibodies, specific T cell responses, or both. A very small proportion (less than 0.01%) of the total lymphocytes are able to bind to a particular antigen, which suggests that only a few cells respond to each antigen.

For the acquired response to "remember" and eliminate a large number of pathogens the immune system must be able to distinguish between many different antigens, and the receptors that recognize antigens must be produced in a huge variety of configurations, in essence one receptor (at least) for each different pathogen that might ever be encountered. Even in the absence of antigen stimulation, a human can produce more than 1 trillion different antibody molecules. Millions of genes would be required to store the genetic information that produces these receptors, but, the entire human genome contains fewer than 25,000 genes.

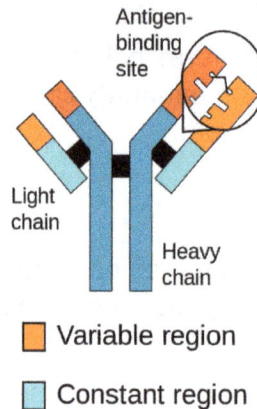

An antibody is made up of two heavy chains and two light chains. The unique variable region allows an antibody to recognize its matching antigen.

Myriad receptors are produced through a process known as clonal selection. According to the clonal selection theory, at birth, an animal randomly generates a vast diversity of lymphocytes (each bearing a unique antigen receptor) from information encoded in a small family of genes. To generate each unique antigen receptor, these genes have undergone a process called V(D)J recombination, or *combinatorial diversification*, in which one gene segment recombines with other gene segments to form a single unique gene. This assembly process generates the enormous diversity of receptors and antibodies, before the body ever encounters antigens, and enables the immune system to respond to an almost unlimited diversity of antigens. Throughout an animal's lifetime, lymphocytes that can react against the antigens an animal actually encounters are selected for action—directed against anything that expresses that antigen.

Note that the innate and acquired portions of the immune system work together, not in spite of each other. The acquired arm, B, and T cells couldn't function without the innate system' input. T cells are useless without antigen-presenting cells to activate them, and B cells are crippled without T cell help. On the other hand, the innate system would likely be overrun with pathogens without the specialized action of the acquired immune response.

Acquired Immunity During Pregnancy

The cornerstone of the immune system is the recognition of "self" versus "non-self". Therefore, the mechanisms that protect the human fetus (which is considered "non-self") from attack by the immune system, are particularly interesting. Although no comprehensive explanation has emerged to explain this mysterious, and often repeated, lack of rejection, two classical reasons may explain how the fetus is tolerated. The first is that the fetus occupies a portion of the body protected by a non-immunological barrier, the uterus, which the immune system does not routinely patrol. The second is that the fetus itself may promote local immunosuppression in the mother, perhaps by a process of active nutrient depletion. A more modern explanation for this induction of tolerance is that specific glycoproteins expressed in the uterus during pregnancy suppress the uterine immune response.

During pregnancy in viviparous mammals (all mammals except Monotremes), endogenous retroviruses (ERVs) are activated and produced in high quantities during the implantation of the embryo. They are currently known to possess immunosuppressive properties, suggesting a role in protecting the embryo from its mother's immune system. Also, viral fusion proteins cause the formation of the placental syncytium to limit exchange of migratory cells between the developing embryo and the body of the mother (something an epithelium can't do sufficiently, as certain blood cells specialize to insert themselves between adjacent epithelial cells). The immunodepressive action was the initial normal behavior of the virus, similar to HIV. The fusion proteins were a way to spread the infection to other cells by simply merging them with the infected one (HIV does this too). It is believed that the ancestors of modern viviparous mammals evolved after an infection by this virus, enabling the fetus to survive the immune system of the mother.

The human genome project found several thousand ERVs classified into 24 families.

Immune Network Theory

A theoretical framework explaining the workings of the acquired immune system is provided by immune network theory, based on interactions between idiotypes (unique molecular features of one clonotype, i.e. the unique set of antigenic determinants of the variable portion of an antibody) and 'anti-idiotypes' (antigen receptors that react with the idiotype as if it were a foreign antigen). This theory, which builds on the existing clonal selection hypothesis and since 1974 has been developed mainly by Niels Jerne and Geoffrey W. Hoffmann, is seen as being relevant to the understanding of the HIV pathogenesis and the search for an HIV vaccine.

Stimulation of Adaptive Immunity

One of the most interesting developments in biomedical science during the past few decades has been elucidation of mechanisms mediating innate immunity. One set of innate immune mechanisms is humoral, such as complement activation. Another set comprises pattern recognition receptors such as Toll-like receptors, which induce the production of interferons and other cytokines increasing resistance of cells such as monocytes to infections. Cytokines produced during innate immune responses are among the activators of adaptive immune responses. Antibodies exert additive or synergistic effects with mechanisms of innate immunity. Unstable HbS clusters Band-3, a major integral red cell protein; antibodies recognize these clusters and accelerate their removal by phagocytic cells. Clustered Band 3 proteins with attached antibodies activate complement, and complement C3 fragments are opsonins recognized by the CR1 complement receptor on phagocytic cells.

A population study has shown that the protective effect of the sickle-cell trait against falciparum malaria involves the augmentation of adaptive as well as innate immune responses to the malaria parasite, illustrating the expected transition from innate to adaptive immunity.

Repeated malaria infections strengthen adaptive immunity and broaden its effects against parasites expressing different surface antigens. By school age most children have developed efficacious adaptive immunity against malaria. These observations raise questions about mechanisms that favor the survival of most children in Africa while allowing some to develop potentially lethal infections.

In malaria, as in other infections, innate immune responses lead into, and stimulate, adaptive immune responses. The genetic control of innate and adaptive immunity is now a large and flourishing discipline.

Humoral and cell-mediated immune responses limit malaria parasite multiplication, and many cytokines contribute to the pathogenesis of malaria as well as to the resolution of infections.

Evolution

The adaptive immune system, which has been best-studied in mammals, originated in a jawed fish approximately 500 million years ago. Most of the molecules, cells, tissues, and associated mechanisms of this system of defense are found in cartilaginous fishes. Lymphocyte receptors, Ig and TCR, are found in all jawed vertebrates. The most ancient Ig class, IgM, is membrane-bound and then secreted upon stimulation of cartilaginous fish B cells. Another isotype, shark IgW, is related to mammalian IgD. TCRs, both α/β and γ/δ, are found in all animals from gnathostomes to mammals. The organization of gene segments that undergo gene rearrangement differs in cartilaginous fishes, which have a cluster form as compared to the translocon form in bony fish to mammals. Like TCR and Ig, the MHC is found only in jawed vertebrates. Genes involved in antigen processing and presentation, as well as the class I and class II genes, are closely linked within the MHC of almost all studied species.

Lymphoid cells can be identified in some pre-vertebrate deuterostomes (i.e., sea urchins). These bind antigen with pattern recognition receptors (PRRs) of the innate immune system. In jawless fishes, two subsets of lymphocytes use variable lymphocyte receptors (VLRs) for antigen binding. Diversity is generated by a cytosine deaminase-mediated rearrangement of LRR-based DNA segments. There is no evidence for the recombination-activating genes (RAGs) that rearrange Ig and TCR gene segments in jawed vertebrates.

The evolution of the AIS, based on Ig, TCR, and MHC molecules, is thought to have arisen from two major evolutionary events: the transfer of the RAG transposon (possibly of viral origin) and two whole genome duplications. Though the molecules of the AIS are well-conserved, they are also rapidly evolving. Yet, a comparative approach finds that many features are quite uniform across taxa. All the major features of the AIS arose early and quickly. Jawless fishes have a different AIS that relies on gene rearrangement to generate diversity but has little else in common with the jawed vertebrate AIS. The innate immune system, which has an important role in AIS activation, is the most important defense system of invertebrates and plants.

Complement System

The complement system is a part of the immune system that enhances (complements) the ability of antibodies and phagocytic cells to clear microbes and damaged cells from an organism, promotes inflammation, and attacks membrane. It is part of the innate immune system, which is not adaptable and does not change over the course of an individual's lifetime. However, it can be recruited and brought into action by the adaptive immune system.

The complement system consists of a number of small proteins found in the blood, in general synthesized by the liver, and normally circulating as inactive precursors (pro-proteins). When stimulated by one of several triggers, proteases in the system cleave specific proteins to release cytokines and initiate an amplifying cascade of further cleavages. The end result of this complement activation or complement fixation cascade is stimulation of phagocytes to clear foreign and damaged material, proxy inflammation to attract additional phagocytes, and activation of the cell-killing membrane attack complex. Over 30 proteins and protein fragments make up the complement system, including serum proteins, serosal proteins, and cell membrane receptors. They account for about 10% of the globulin fraction of blood serum and can serve as opsonins.

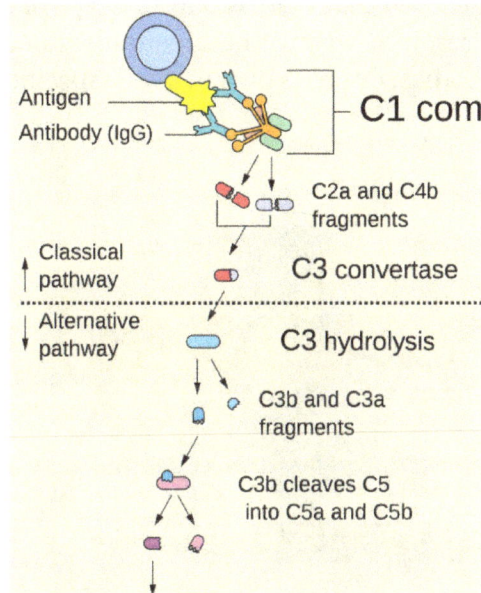

Scheme of the complement system

Three biochemical pathways activate the complement system: the classical complement pathway, the alternative complement pathway, and the lectin pathway.

History

In 1888, George Nuttall found that sheep blood serum had mild killing activity against the bacterium that causes anthrax. The killing activity disappeared when he heated the blood. In 1891, Hans Ernst August Buchner, noting the same property of blood in his experiments, named the killing property "alexin," which means "to ward off" in Greek. By 1884, several laboratories had demonstrated that serum from guinea pigs that had recovered from cholera killed the cholera bacterium in vitro. Heating the serum destroyed its killing activity. Nevertheless, the heat-inactivated serum, when injected into guinea pigs exposed to the cholera bacteria, maintained its ability to protect the animals from illness. Jules Bordet, a young Belgian scientist in Paris at the Pasteur Institute, concluded that this principle has two components, one that maintained a "sensitizing" effect after being heated and one (alexin) whose toxic effect was lost after being heated. The heat-stable component was responsible for immunity against specific microorganisms, whereas the heat-sensitive component was responsible for the non-specific antimicrobial activity conferred by all normal sera. In 1899, Paul Ehrlich renamed the heat-sensitive component "complement."

Ehrlich introduced the term "complement" as part of his larger theory of the immune system. According to this theory, the immune system consists of cells that have specific receptors on their surface to recognize antigens. Upon immunisation with an antigen, more of these receptors are formed, and they are then shed from the cells to circulate in the blood. These receptors, which we now call "antibodies," were called by Ehrlich "amboceptors" to emphasise their bifunctional binding capacity: They recognise and bind to a specific antigen, but they also recognise and bind to the heat-labile antimicrobial component of fresh serum. Ehrlich, therefore, named this heat-labile component "complement," because it is something in the blood that "complements" the cells of the immune system. Ehrlich believed that each antigen-specific amboceptor has its own specific complement, whereas Bordet believed that there is only one type of complement. In the early 20th century, this controversy was resolved when it became understood that complement can act in combination with specific antibodies, or on its own in a non-specific way.

Functions

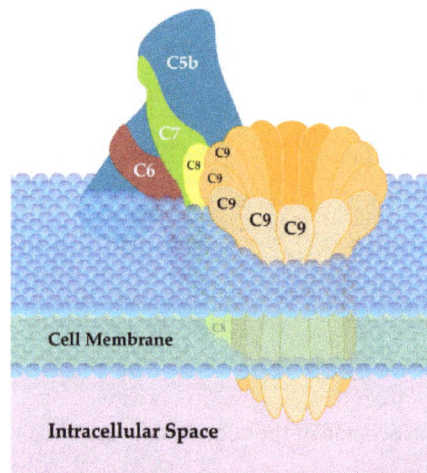

Membrane Attack Complex (Terminal Complement Complex C5b-9)

Complement triggers the following immune functions:

1. Phagocytosis – by opsonizing antigens. C3b has most important opsonizing activity

2. Inflammation – by attracting macrophages and neutrophils

3. Membrane attack – by rupturing membranes of foreign cells

Overview

Most of the proteins and glycoproteins that constitute the complement system are synthesized by hepatocytes. But significant amounts are also produced by tissue macrophages, blood monocytes, and epithelial cells of the genitourinary tract and gastrointestinal tract. The three pathways of activation all generate homologous variants of the protease C3-convertase. The classical complement pathway typically requires antigen-antibody complexes (immune complexes) for activation (specific immune response), whereas the alternative pathway can be activated by C3 hydrolysis, foreign material, pathogens, or damaged cells. The mannose-binding lectin pathway can be activated by C3 hydrolysis or antigens without the presence of antibodies (non-specific immune response).

In all three pathways, C3-convertase cleaves and activates component C3, creating C3a and C3b, and causes a cascade of further cleavage and activation events. C3b binds to the surface of pathogens, leading to greater internalization by phagocytic cells by opsonization.

In the alternative pathway, C3b binds to Factor B. Factor D releases Factor Ba from Factor B bound to C3b. The complex of C3b(2)Bb is a protease which cleaves C5 into C5b and C5a. C5 convertase is also formed by the Classical Pathway when C3b binds C4b and C2a. C5a is an important chemotactic protein, helping recruit inflammatory cells. C3a is the precursor of an important cytokine (adipokine) named ASP (although this is not universally accepted) and is usually rapidly cleaved by carboxypeptidase B. Both C3a and C5a have anaphylatoxin activity, directly triggering degranulation of mast cells as well as increasing vascular permeability and smooth muscle contraction. C5b initiates the membrane attack pathway, which results in the membrane attack complex (MAC), consisting of C5b, C6, C7, C8, and polymeric C9. MAC is the cytolytic endproduct of the complement cascade; it forms a transmembrane channel, which causes osmotic lysis of the target cell. Kupffer cells and other macrophage cell types help clear complement-coated pathogens. As part of the innate immune system, elements of the complement cascade can be found in species earlier than vertebrates; most recently in the protostome horseshoe crab species, putting the origins of the system back further than was previously thought.

Classical Pathway

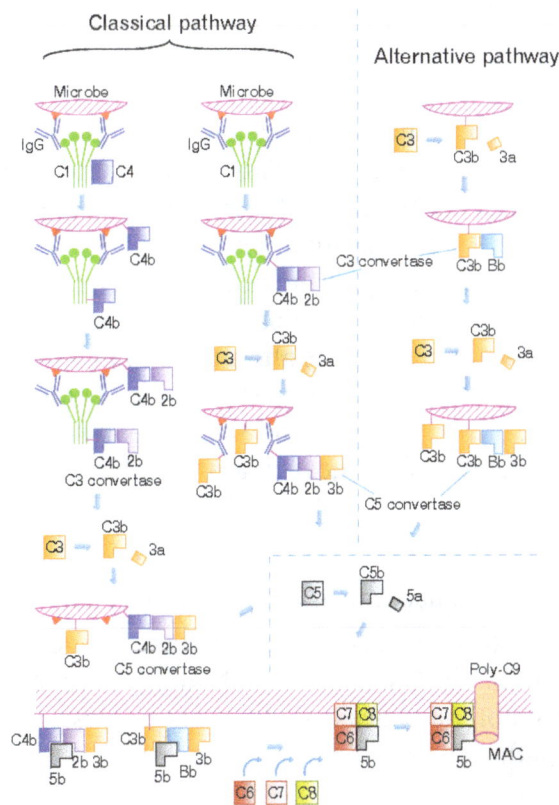

Figure 2. The classical and alternative complement pathways

The classical pathway is triggered by activation of the C1-complex. The C1-complex is composed of 1 molecule of C1q, 2 molecules of C1r and 2 molecules of C1s, or *C1qr₂s₂*. This occurs when C1q binds to IgM or IgG complexed with antigens. A single pentameric IgM can initiate the pathway, while several, ideally six, IgGs are needed. This also occurs when C1q binds directly to the surface of the pathogen. Such binding leads to conformational changes in the C1q molecule, which leads to the activation of two C1r molecules. C1r is a serine protease. They then cleave C1s (another serine protease). The $C1r^2s^2$ component now splits C4 and then C2, producing C4a, C4b, C2a, and C2b. C4b and C2a bind to form the classical pathway C3-convertase (C4b2a complex), which promotes cleavage of C3 into C3a and C3b; C3b later joins with C4b2a, the C3 convertase) to make C5 convertase (C4b2a3b complex).

Alternative Pathway

The alternative pathway is continuously activated at a low level, analogous to a car engine at idle, as a result of spontaneous C3 hydrolysis due to the breakdown of the internal thioester bond (C3 is mildly unstable in aqueous environment). The alternative pathway does not rely on pathogen-binding antibodies like the other pathways. C3b that is generated from C3 by a C3 convertase enzyme complex in the fluid phase is rapidly inactivated by factor H and factor I, as is the C3b-like C3 that is the product of spontaneous cleavage of the internal thioester. In contrast, when the internal thioester of C3 reacts with a hydroxyl or amino group of a molecule on the surface of a cell or pathogen, the C3b that is now covalently bound to the surface is protected from factor H-mediated inactivation. The surface-bound C3b may now bind factor B to form C3bB. This complex in the presence of factor D will be cleaved into Ba and Bb. Bb will remain associated with C3b to form C3bBb, which is the alternative pathway C3 convertase.

The C3bBb complex is stabilized by binding oligomers of factor P (Properdin). The stabilized C3 convertase, C3bBbP, then acts enzymatically to cleave much more C3, some of which becomes covalently attached to the same surface as C3b. This newly bound C3b recruits more B, D and P activity and greatly amplifies the complement activation. When complement is activated on a cell surface, the activation is limited by endogenous complement regulatory proteins, which include CD35, CD46, CD55 and CD59, depending on the cell. Pathogens, in general, don't have complement regulatory proteins (there are many exceptions, which reflect adaptation of microbial pathogens to vertebrate immune defenses). Thus, the alternative complement pathway is able to distinguish self from non-self on the basis of the surface expression of complement regulatory proteins. Host cells don't accumulate cell surface C3b (and the proteolytic fragment of C3b called iC3b) because this is prevented by the complement regulatory proteins, while foreign cells, pathogens and abnormal surfaces may be heavily decorated with C3b and iC3b. Accordingly, the alternative complement pathway is one element of innate immunity.

Once the alternative C3 convertase enzyme is formed on a pathogen or cell surface, it may bind covalently another C3b, to form C3bBbC3bP, the C5 convertase. This enzyme then cleaves C5 to C5a, a potent anaphylatoxin, and C5b. The C5b then recruits and assembles C6, C7, C8 and multiple C9 molecules to assemble the membrane attack complex. This creates a hole or pore in the membrane that can kill or damage the pathogen or cell.

Lectin Pathway

The lectin pathway is homologous to the classical pathway, but with the opsonin, mannose-binding lectin (MBL), and ficolins, instead of C1q. This pathway is activated by binding of MBL to mannose residues on the pathogen surface, which activates the MBL-associated serine proteases, MASP-1, and MASP-2 (very similar to C1r and C1s, respectively), which can then split C4 into C4a and C4b and C2 into C2a and C2b. C4b and C2a then bind together to form the classical C3-convertase, as in the classical pathway. Ficolins are homologous to MBL and function via MASP in a similar way. Several single-nucleotide polymorphisms have been described in M-ficolin in humans, with effect on ligand-binding ability and serum levels. Historically, the larger fragment of C2 was named C2a, but it is now referred to as C2b. In invertebrates without an adaptive immune system, ficolins are expanded and their binding specificities diversified to compensate for the lack of pathogen-specific recognition molecules.

Complement Protein Fragment Nomenclature

Immunology textbooks have used different naming assignments for the smaller and larger fragments of C2 as C2a and C2b. The preferred assignment appears to be that the smaller fragment be designated as C2a: as early as 1994, a well known textbook recommended that the larger fragment of C2 should be designated C2b. However, this was amplified in their 1999 4th edition, to say that: "It is also useful to be aware that the larger active fragment of C2 was originally designated C2a, and is still called that in some texts and research papers. Here, for consistency, we shall call all large fragments of complement b, so the larger fragment of C2 will be designated C2b. In the classical and lectin pathways the C3 convertase enzyme is formed from membrane-bound C4b with C2b."

This nomenclature is used in another literature: "(Note that, in older texts, the smaller fragment is often called C2b, and the larger one is called C2a for historical reason.)" The assignment is mixed in the latter literature, though. Some sources designate the larger and smaller fragments as C2a and C2b respectively while other sources apply the converse. However, due to the widely established convention, C2b here is the larger fragment, which, in the classical pathway, forms C4b2b (classically C4b2a). It may be noteworthy that, in a series of editions of Janeway's book, 1st to 7th, in the latest edition they withdraw the stance to indicate the larger fragment of C2 as C2b.

Viral Inhibition

Fixation of the C1a protein on viral surfaces has also been shown to enhance neutralization of viral pathogens.

Activation of Complements by Antigen-Associated Antibody

In the classical pathway, C1 binds with its C1q subunits to Fc fragments (made of CH2 region) of IgG or IgM, which has formed a complex with antigens. C4b and C3b are also able to bind to antigen-associated IgG or IgM, to its Fc portion.

Such immunoglobulin-mediated binding of the complement may be interpreted as that the complement uses the ability of the immunoglobulin to detect and bind to non-self antigens as its guiding stick. The complement itself is able to bind non-self pathogens after detecting their pathogen-associated molecular patterns (PAMPs), however, utilizing specificity of antibody, complements are able to detect non-self enemies much more specifically.

Figure 2 shows the classical and the alternative pathways with the late steps of complement activation schematically. Some components have a variety of binding sites. In the classical pathway C4 binds to Ig-associated C1q and C1r²s² enzyme cleave C4 to C4b and 4a. C4b binds to C1q, antigen-associated Ig (specifically to its Fc portion), and even to the microbe surface. C3b binds to antigen-associated Ig and to the microbe surface. Ability of C3b to bind to antigen-associated Ig would work effectively against antigen-antibody immune complexes to make them soluble. In the figure, C2b refers to the larger of the C2 fragments.

Regulation

The complement system has the potential to be extremely damaging to host tissues, meaning its activation must be tightly regulated. The complement system is regulated by complement control proteins, which are present at a higher concentration in the blood plasma than the complement proteins themselves. Some complement control proteins are present on the membranes of self-cells preventing them from being targeted by complement. One example is CD59, also known as protectin, which inhibits C9 polymerisation during the formation of the membrane attack complex. The classical pathway is inhibited by C1-inhibitor, which binds to C1 to prevent its activation.

C3-convertase can be inhibited by Decay accelerating factor (DAF), which is bound to erythrocyte plasma membranes via a GPI anchor.

Role in Disease

Complement Deficiency

It is thought that the complement system might play a role in many diseases with an immune component, such as Barraquer-Simons Syndrome, asthma, lupus erythematosus, glomerulonephritis, various forms of arthritis, autoimmune heart disease, multiple sclerosis, inflammatory bowel disease, paroxysmal nocturnal hemoglobinuria, atypical hemolytic uremic syndrome and ischemia-reperfusion injuries, and rejection of transplanted organs.

The complement system is also becoming increasingly implicated in diseases of the central nervous system such as Alzheimer's disease and other neurodegenerative conditions such as spinal cord injuries.

Deficiencies of the terminal pathway predispose to both autoimmune disease and infections (particularly Neisseria meningitidis, due to the role that the membrane attack complex ("MAC") plays in attacking Gram-negative bacteria).

Infections with N. meningitidis and N. gonorrhoeae are the only conditions known to be associated with deficiencies in the MAC components of complement. 40-50% of those with MAC deficiencies experience recurrent infections with N. meningitidis.

Deficiencies in Complement Regulators

Mutations in the complement regulators factor H and membrane cofactor protein have been associated with atypical hemolytic uremic syndrome. Moreover, a common single nucleotide polymorphism in factor H (Y402H) has been associated with the common eye disease age-related macular degeneration. Polymorphisms of complement component 3, complement factor B, and complement factor I, as well as deletion of complement factor H-related 3 and complement factor H-related 1 also affect a person's risk of developing age-related macular degeneration. Both of these disorders are currently thought to be due to aberrant complement activation on the surface of host cells.

Mutations in the C1 inhibitor gene can cause hereditary angioedema, a genetic condition resulting from reduced regulation of bradykinin by C1-INH.

Paroxysmal nocturnal hemoglobinuria is caused by complement breakdown of RBCs due to an inability to make GPI. Thus the RBCs are not protected by GPI anchored proteins such as DAF.

Diagnostic Tools

Diagnostic tools to measure complement activity include the total complement activity test.

The presence or absence of complement fixation upon a challenge can indicate whether particular antigens or antibodies are present in the blood. This is the principle of the complement fixation test.

Modulation by Infections

Recent research has suggested that the complement system is manipulated during HIV/AIDS to further damage the body.

References

- Janeway, Charles; Paul Travers; Mark Walport; Mark Shlomchik (2001). Immunobiology; Fifth Edition. New York and London: Garland Science. ISBN 0-8153-4101-6..

- Janeway, C.A.; Travers, P.; Walport, M.; Shlomchik, M.J. (2005). Immunobiology. (6th ed.). Garland Science. ISBN 0-443-07310-4.

- Dorland WAN (editor) (2003). Dorland's Illustrated Medical Dictionary (30th ed.). W.B. Saunders. ISBN 0-7216-0146-4.

- Alberts, B.; Johnson, A.; Lewis, J.; Raff, M.; Roberts, K.; Walters, P. (2002). Molecular Biology of the Cell (4th ed.). New York and London: Garland Science. ISBN 0-8153-3218-1.

- Janeway, C.A.; Travers, P.; Walport, M.; Shlomchik, M.J. (2001). Immunobiology (5th ed.). New York and London: Garland Science. ISBN 0-8153-4101-6..

- Janeway, C.A.; Travers, P.; Walport, M.; Shlomchik, M.J. (2005). Immunobiology. (6th ed.). Garland Science. ISBN 0-443-07310-4.

- Flajnik, MF; Kasahara, M (Jan 2010). "Origin and evolution of the adaptive immune system: genetic events and selective pressures.". Nature Reviews Genetics. 11 (1): 47–59. doi:10.1038/nrg2703. PMC 3805090. PMID 19997068.

Lymphatic System: Organs and Tissues

The lymphatic system is a part of the circulatory system and a vital part of the immune system consists of a network of lymphatic vessels that carry a clear fluid called lymph toward the heart. This chapter studies the lymphatic system and the organs and tissues that comprise it. Also in this chapter is a description of the role of lymph in the immune function of the body.

Lymphatic System

The lymphatic system is part of the circulatory system and a vital part of the immune system, comprising a network of lymphatic vessels that carry a clear fluid called lymph (from Latin, *lympha* meaning *water*) directionally towards the heart. The lymphatic system was first described in the seventeenth century independently by Olaus Rudbeck and Thomas Bartholin. Unlike the cardiovascular system, the lymphatic system is not a closed system. The human circulatory system processes an average of 20 liters of blood per day through capillary filtration, which removes plasma while leaving the blood cells. Roughly 17 litres of the filtered plasma are reabsorbed directly into the blood vessels, while the remaining three litres remain in the interstitial fluid. One of the main functions of the lymph system is to provide an accessory return route to the blood for the surplus three litres.

The other main function is that of defense in the immune system. Lymph is very similar to blood plasma: it contains lymphocytes and other white blood cells. It also contains waste products and cellular debris together with bacteria and proteins. Associated organs composed of lymphoid tissue are the sites of lymphocyte production. Lymphocytes are concentrated in the lymph nodes. The spleen and the thymus are also lymphoid organs of the immune system. The tonsils are lymphoid organs that are also associated with the digestive system. Lymphoid tissues contain lymphocytes, and also contain other types of cells for support. The system also includes all the structures dedicated to the circulation and production of lymphocytes (the primary cellular component of lymph), which also includes the bone marrow, and the lymphoid tissue associated with the digestive system.

The blood does not come into direct contact with the parenchymal cells and tissues in the body (except in case of an injury causing rupture of one or more blood vessels), but constituents of the blood first exit the microvascular exchange blood vessels to become interstitial fluid, which comes into contact with the parenchymal cells of the body. Lymph is the fluid that is formed when interstitial fluid enters the initial lymphatic vessels of the lymphatic system. The lymph is then moved along the lymphatic vessel network by either intrinsic contractions of the lymphatic passages or by extrinsic compression of the lymphatic vessels via external tissue forces (e.g., the contractions of skeletal muscles), or by lymph hearts in some animals. The organization of lymph nodes and drainage follows the organization of the body into external and internal regions; therefore, the lymphatic drainage of the head, limbs, and body cavity walls follows an external route, and the lymphatic drainage of the thorax, abdomen, and pelvic cavities follows an internal route. Eventually, the lymph vessels empty into the lymphatic ducts, which drain into one of the two subclavian veins, near their junction with the internal jugular veins.

Structure

The lymphatic system consists of lymphatic organs, a conducting network of lymphatic vessels, and the circulating lymph.

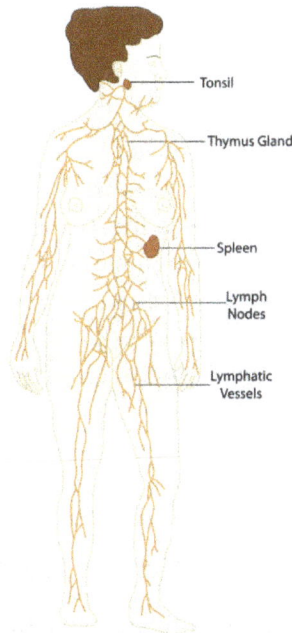

Lymphatic system

The thymus and the bone marrow constitute the primary lymphoid organs involved in the production and early clonal selection of lymphocyte tissues. Bone marrow is responsible for both the creation of T cells and the production and maturation of B cells. From the bone marrow, B cells immediately join the circulatory system and travel to secondary lymphoid organs in search of pathogens. T cells, on the other hand, travel from the bone marrow to the thymus, where they develop further. Mature T cells join B cells in search of pathogens. The other 95% of T cells begin a process of apoptosis (programmed cell death).

The central or primary lymphoid organs generate lymphocytes from immature progenitor cells.

Secondary or peripheral lymphoid organs, which include lymph nodes and the spleen, maintain mature naive lymphocytes and initiate an adaptive immune response. The peripheral lymphoid organs are the sites of lymphocyte activation by antigens. Activation leads to clonal expansion and affinity maturation. Mature lymphocytes recirculate between the blood and the peripheral lymphoid organs until they encounter their specific antigen.

Secondary lymphoid tissue provides the environment for the foreign or altered native molecules (antigens) to interact with the lymphocytes. It is exemplified by the lymph nodes, and the lymphoid follicles in tonsils, Peyer's patches, spleen, adenoids, skin, etc. that are associated with the mucosa-associated lymphoid tissue (MALT).

In the gastrointestinal wall the appendix has mucosa resembling that of the colon, but here it is heavily infiltrated with lymphocytes.

The tertiary lymphoid tissue typically contains far fewer lymphocytes, and assumes an immune role only when challenged with antigens that result in inflammation. It achieves this by importing the lymphocytes from blood and lymph.)

Lymphoid Tissue

Thymus

The thymus is a primary lymphoid organ and the site of maturation for T cells, the lymphocytes of the adaptive immune system. The thymus increases in size from birth in response to postnatal antigen stimulation, then to puberty and regresses thereafter. The loss or lack of the thymus results in severe immunodeficiency and subsequent high susceptibility to infection. In most species, the thymus consists of lobules divided by septa which are made up of epithelium and is therefore an epithelial organ. T cells mature from thymocytes, proliferate and undergo selection process in the thymic cortex before entering the medulla to interact with epithelial cells.

Spleen

The main functions of the spleen are:

1. to produce immune response against blood-borne antigens

2. to remove particulate matter and aged blood cells, mainly erythrocytes

3. to produce blood cells during fetal life

The spleen synthesizes antibodies in its white pulp and removes antibody-coated bacteria and antibody-coated blood cells by way of blood and lymph node circulation. A study published in 2009 using mice found that the spleen contains, in its reserve, half of the body's monocytes within the red pulp. These monocytes, upon moving to injured tissue (such as the heart), turn into dendritic cells and macrophages while promoting tissue healing. The spleen is a center of activity of the mononuclear phagocyte system and can be considered analogous to a large lymph node, as its absence causes a predisposition to certain infections.

Like the thymus, the spleen has only efferent lymphatic vessels. Both the short gastric arteries and the splenic artery supply it with blood.

The germinal centers are supplied by arterioles called *penicilliary radicles*.

Up to the fifth month of prenatal development the spleen creates red blood cells. After birth the bone marrow is solely responsible for hematopoiesis. As a major lymphoid organ and a central player in the reticuloendothelial system, the spleen retains the ability to produce lymphocytes. The spleen stores red blood cells and lymphocytes. It can store enough blood cells to help in an emergency. Up to 25% of lymphocytes can be stored at any one time.

Lymph Nodes

A lymph node is an organized collection of lymphoid tissue, through which the lymph passes on its way back to the blood. Lymph nodes are located at intervals along the lymphatic system. Several afferent lymph vessels bring in lymph, which percolates through the substance of the lymph node, and is then drained out by an efferent lymph vessel. There are between five and six hundred lymph nodes in the human body, many of which are grouped in clusters in different regions as in the underarm and abdominal areas. Lymph node clusters are commonly found at the base of limbs (groin, armpits) and in the neck, where lymph is collected from regions of the body likely to sustain pathogen contamination from injuries.

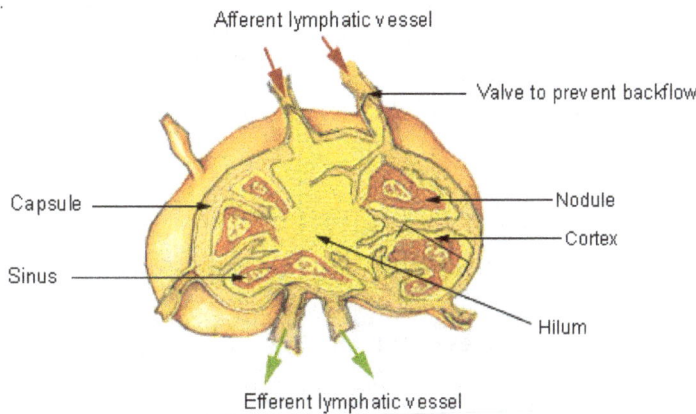

A lymph node showing afferent and efferent lymphatic vessels

The substance of a lymph node consists of lymphoid follicles in an outer portion called the cortex. The inner portion of the node is called the medulla, which is surrounded by the cortex on all sides except for a portion known as the hilum. The hilum presents as a depression on the surface of the lymph node, causing the otherwise spherical lymph node to be bean-shaped or ovoid. The efferent lymph vessel directly emerges from the lymph node at the hilum. The arteries and veins supplying the lymph node with blood enter and exit through the hilum.

The region of the lymph node called the paracortex immediately surrounds the medulla. Unlike the cortex, which has mostly immature T cells, or thymocytes, the paracortex has a mixture of immature and mature T cells. Lymphocytes enter the lymph nodes through specialised high endothelial venules found in the paracortex.

A lymph follicle is a dense collection of lymphocytes, the number, size and configuration of which change in accordance with the functional state of the lymph node. For example, the follicles expand significantly when encountering a foreign antigen. The selection of B cells, or *B lymphocytes*, occurs in the germinal center of the lymph nodes.

Lymph nodes are particularly numerous in the mediastinum in the chest, neck, pelvis, axilla, inguinal region, and in association with the blood vessels of the intestines.

Other Lymphoid Tissue

Lymphoid tissue associated with the lymphatic system is concerned with immune functions in defending the body against infections and the spread of tumors. It consists of connective tissue formed of reticular fibers, with various types of leukocytes, (white blood cells), mostly lymphocytes enmeshed in it, through which the lymph passes. Regions of the lymphoid tissue that are densely packed with lymphocytes are known as *lymphoid follicles.* Lymphoid tissue can either be structurally well organized as lymph nodes or may consist of loosely organized lymphoid follicles known as the mucosa-associated lymphoid tissue.

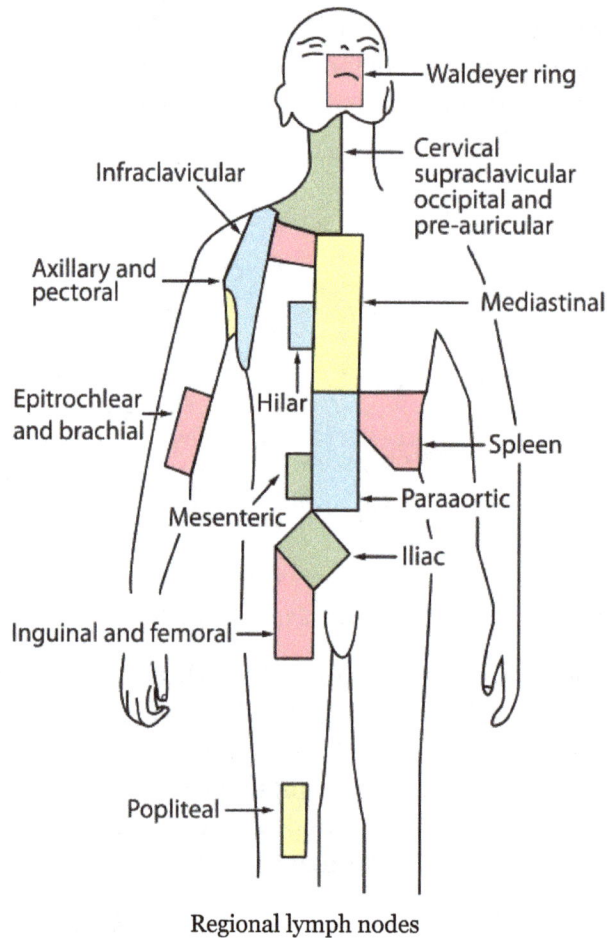

Regional lymph nodes

The central nervous system also has lymphatic vessels, as discovered by University of Virgina Researchers. The search for T-cell gateways into and out of the meninges uncovered functional lymphatic vessels lining the dural sinuses, anatomically integrated into the membrane surrounding the brain.

Lymphatics

The lymphatic vessels, also called lymph vessels, conduct lymph between different parts of the body. They include the tubular vessels of the lymph capillaries, and the larger collecting vessels–the right lymphatic duct and the thoracic duct (the left lymphatic duct). The lymph capillaries are mainly responsible for the absorption of interstitial fluid from the tissues, while lymph vessels propel the absorbed fluid forward into the larger collecting ducts, where it ultimately returns to the bloodstream via one of the subclavian veins. These vessels are also called the lymphatic channels or simply *lymphatics.*

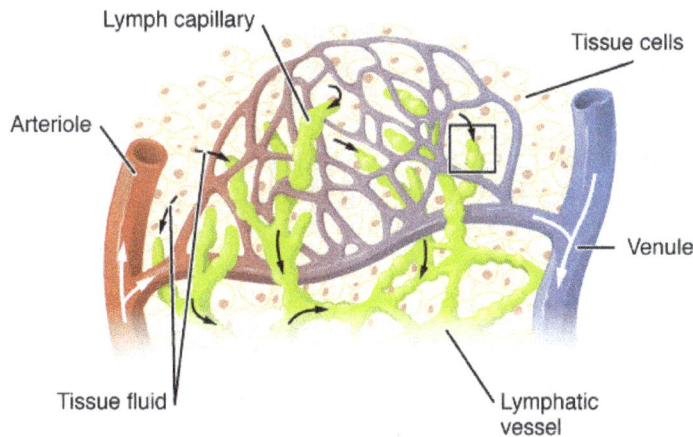

Lymph capillaries in the tissue spaces

The lymphatics are responsible for maintaining the balance of the body fluids. Its network of capillaries and collecting lymphatic vessels work to efficiently drain and transport extravasated fluid, along with proteins and antigens, back to the circulatory system. Numerous intraluminal valves in the vessels ensure a unidirectional flow of lymph without reflux. Two valve systems are used to achieve this one directional flow—a primary and a secondary valve system. The capillaries are blind-ended, and the valves at the ends of capillaries use specialised junctions together with anchoring filaments to allow a unidirectional flow to the primary vessels. The collecting lymphatics, however, act to propel the lymph by the combined actions of the intraluminal valves and lymphatic muscle cells.

Development

Lymphatic tissues begin to develop by the end of the fifth week of embryonic development. Lymphatic vessels develop from lymph sacs that arise from developing veins, which are derived from mesoderm.

The first lymph sacs to appear are the paired jugular lymph sacs at the junction of the internal jugular and subclavian veins. From the jugular lymph sacs, lymphatic capillary plexuses spread to the thorax, upper limbs, neck and head. Some of the plexuses enlarge and form lymphatic vessels in their respective regions. Each jugular lymph sac retains at least one connection with its jugular vein, the left one developing into the superior portion of the thoracic duct.

The next lymph sac to appear is the unpaired retroperitoneal lymph sac at the root of the mesentery of the intestine. It develops from the primitive vena cava and mesonephric veins. Capillary plexuses and lymphatic vessels spread from the retroperitoneal lymph sac to the abdominal viscera and diaphragm. The sac establishes connections with the cisterna chyli but loses its connections with neighboring veins.

The last of the lymph sacs, the paired posterior lymph sacs, develop from the iliac veins. The posterior lymph sacs produce capillary plexuses and lymphatic vessels of the abdominal wall, pelvic region, and lower limbs. The posterior lymph sacs join the cisterna chyli and lose their connections with adjacent veins.

With the exception of the anterior part of the sac from which the cisterna chyli develops, all lymph sacs become invaded by mesenchymal cells and are converted into groups of lymph nodes.

The spleen develops from mesenchymal cells between layers of the dorsal mesentery of the stomach. The thymus arises as an outgrowth of the third pharyngeal pouch.

Function

The lymphatic system has multiple interrelated functions:

- It is responsible for the removal of interstitial fluid from tissues
- It absorbs and transports fatty acids and fats as chyle from the digestive system
- It transports white blood cells to and from the lymph nodes into the bones
- The lymph transports antigen-presenting cells, such as dendritic cells, to the lymph nodes where an immune response is stimulated.

Function of The Fatty Acid Transport System

Lymph vessels called lacteals are present in the lining of the gastrointestinal tract, predominantly in the small intestine. While most other nutrients absorbed by the small intestine are passed on to the portal venous system to drain via the portal vein into the liver for processing, fats (lipids) are passed on to the lymphatic system to be transported to the blood circulation via the thoracic duct. (There are exceptions, for example medium-chain triglycerides are fatty acid esters of glycerol that passively diffuse from the GI tract to the portal system.) The enriched lymph originating in the lymphatics of the small intestine is called chyle. The nutrients that are released to the circulatory system are processed by the liver, having passed through the systemic circulation.

Immune Function

The lymphatic system plays a major role in body's immune system, as the primary site for cells relating to adaptive immune system including T-cells and B-cells. Cells in the lymphatic system react to antigens presented or found by the cells directly or by other dendritic cells. When an antigen is recognised, an immunological cascade begins involving the activation and recruitment of more and more cells, the production of antibodies and cytokines and the recruitment of other immunological cells such as macrophages.

Clinical Significance

The study of lymphatic drainage of various organs is important in the diagnosis, prognosis, and treatment of cancer. The lymphatic system, because of its closeness to many tissues of the body, is responsible for carrying cancerous cells between the various parts of the body in a process called metastasis. The intervening lymph nodes can trap the cancer cells. If they are not successful in destroying the cancer cells the nodes may become sites of secondary tumors.

Lymphadenopathy

Lymphadenopathy refers to one or more enlarged lymph nodes. Small groups or individually enlarged lymph nodes are generally *reactive* in response to infection or inflammation. This is called *local* lymphadenopathy. When many lymph nodes in different areas of the body are involved, this is called *generalised* lymphadenopathy. Generalised lymphadenopathy may be caused by infections such as infectious mononucleosis, tuberculosis and HIV, connective tissue diseases such as SLE and rheumatoid arthritis, and cancers, including both cancers of tissue within lymph nodes, discussed below, and metastasis of cancerous cells from other parts of the body, that have arrived via the lymphatic system.

Lymphedema

Lymphedema is the swelling caused by the accumulation of lymph, which may occur if the lymphatic system is damaged or has malformations. It usually affects limbs, though the face, neck and abdomen may also be affected. In an extreme state, called elephantiasis, the edema progresses to the extent that the skin becomes thick with an appearance similar to the skin on elephant limbs.

Causes are unknown in most cases, but sometimes there is a previous history of severe infection, usually caused by a parasitic disease, such as lymphatic filariasis.

Lymphangiomatosis is a disease involving multiple cysts or lesions formed from lymphatic vessels.

Lymphedema can also occur after surgical removal of cancerous lymph nodes in the armpit (causing the arm to swell due to poor lymphatic drainage) or groin (causing swelling of the leg). Treatment is by massage, and is not permanent.

Cancer

Cancer of the lymphatic system can be primary or secondary. Lymphoma refers to cancer that arises from lymphatic tissue. Lymphoid leukemias and lymphomas are now considered to be tumors of the same type of cell lineage. They are called "leukemia" when in the blood or marrow and "lymphoma" when in lymphatic tissue. They are grouped together under the name "lymphoid malignancy".

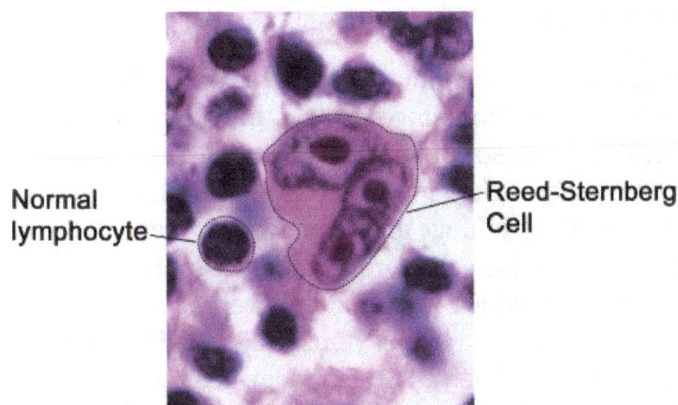

Reed–Sternberg cells.

Lymphoma is generally considered as either Hodgkin lymphoma or non-Hodgkin lymphoma. Hodgkin lymphoma is characterised by a particular type of cell, called a Reed–Sternberg cell, visible under microscope. It is associated with past infection with the Epstein-Barr Virus, and generally causes a painless "rubbery" lymphadenopathy. It is staged, using Ann Arbor staging. Chemotherapy generally involves the ABVD and may also involve radiotherapy. Non-Hodgkin lymphoma is a cancer characterised by increased proliferation of B-cells or T-cells, generally occurs in an older age group than Hodgkin lymphoma. It is treated according to whether it is *high-grade* or *low-grade*, and carries a poorer prognosis than Hodgkin lymphoma.

Lymphangiosarcoma is a malignant soft tissue tumor, whereas lymphangioma is a benign tumor occurring frequently in association with Turner syndrome. Lymphangioleiomyomatosis is a benign tumor of the smooth muscles of the lymphatics that occurs in the lungs.

Other

- Lymphangitis
- Kikuchi disease
- Chylothorax
- Castleman's disease
- Lymphatic filariasis
- Solitary lymphatic nodule

History

Hippocrates, in 5th century BC, was one of the first people to mention the lymphatic system. In his work *On Joints*, he briefly mentioned the lymph nodes in one sentence. Rufus of Ephesus, a Roman physician, identified the axillary, inguinal and mesenteric lymph nodes as well as the thymus during the 1st to 2nd century AD. The first mention of lymphatic vessels was in 3rd century BC by Herophilos, a Greek anatomist living in Alexandria, who incorrectly concluded that the "absorptive veins of the lymphatics," by which he meant the lacteals (lymph vessels of the intestines), drained into the hepatic portal veins, and thus into the liver. The findings of Ruphus and Herophilos were further propagated by the Greek physician Galen, who described the lacteals and mesenteric lymph nodes which he observed in his dissection of apes and pigs in the 2nd century AD.

In the mid 16th century, Gabriele Falloppio (discoverer of the fallopian tubes), described what are now known as the lacteals as "coursing over the intestines full of yellow matter." In about 1563 Bartolomeo Eustachi, a professor of anatomy, described the thoracic duct in horses as *vena alba thoracis*. The next breakthrough came when in 1622 a physician, Gaspare Aselli, identified lymphatic vessels of the intestines in dogs and termed them *venae alba et lacteae*, which is now known as simply the lacteals. The lacteals were termed the fourth kind of vessels (the other three being the artery, vein and nerve, which was then believed to be a type of vessel), and disproved Galen's assertion that chyle was carried by the veins. But, he still believed that the lacteals carried the chyle to the liver (as taught by Galen). He also identified the thoracic duct but failed to notice its connection with the lacteals. This connection was established by Jean Pecquet in 1651, who found a white fluid mixing with blood in a dog's heart. He suspected that fluid to be chyle as its flow increased when abdominal pressure was applied. He traced this fluid to the thoracic duct, which he then followed to a chyle-filled sac he called the *chyli receptaculum,* which is now known as the cisternae chyli; further investigations led him to find that lacteals' contents enter the venous system via the thoracic duct. Thus, it was proven convincingly that the lacteals did not terminate in the liver, thus disproving Galen's second idea: that the chyle flowed to the liver. Johann Veslingius drew the earliest sketches of the lacteals in humans in 1647.

The idea that blood recirculates through the body rather than being produced anew by the liver and the heart was first accepted as a result of works of William Harvey—a work he published in 1628. In 1652, Olaus Rudbeck (1630–1702), a Swede, discovered certain transparent vessels in the liver that contained clear fluid (and not white), and thus named them *hepatico-aqueous vessels.* He also learned that they emptied into the thoracic duct, and that they had valves. He announced his findings in the court of Queen Christina of Sweden, but did not publish his findings for a year, and in the interim similar findings were published by Thomas Bartholin, who additionally published that such vessels are present everywhere in the body, not just in the liver. He is also the one to have named them "lymphatic vessels." This had resulted in a bitter dispute between one of Bartholin's pupils, Martin Bogdan, and Rudbeck, whom he accused of plagiarism.

Galen's ideas prevailed in medicine until the 17th century. It was believed that blood was produced by the liver from chyle contaminated with ailments by the intestine and stomach, to which various spirits were added by other organs, and that this blood was consumed by all the organs of the body. This theory required that the blood be consumed and produced many times over. Even in the 17th century, his ideas were defended by some physicians.

Alexander Monro, of the University of Edinburgh Medical School, was the first to describe the function of the lymphatic system in detail.

"Claude Galien". Lithograph by Pierre Roche Vigneron. (Paris: Lith de Gregoire et Deneux, ca. 1865)

Etymology

The adjective used for the lymph-transporting system is "lymphatic." The adjective used for the tissues where lymphocytes are formed is "lymphoid."

Lymphatic comes from the Latin word lymphaticus, meaning "connected to water."

Thymus

The thymus is a specialized primary lymphoid organ of the immune system. Within the thymus, T cells or T lymphocytes mature. T cells are critical to the adaptive immune system, where the body adapts specifically to foreign invaders. The thymus is composed of two identical lobes and is located anatomically in the anterior superior mediastinum, in front of the heart and behind the sternum. Histologically, each lobe of the thymus can be divided into a central medulla and a peripheral cortex which is surrounded by an outer capsule. The cortex and medulla play different roles in the development of T-cells. Cells in the thymus can be divided into thymic stromal cells and cells of hematopoietic origin (derived from bone marrow resident hematopoietic stem cells). Developing T-cells are referred to as thymocytes and are of hematopoietic origin. Stromal cells include epithelial cells of the thymic cortex and medulla, and dendritic cells.

The thymus provides an inductive environment for development of T cells from hematopoietic progenitor cells. In addition, thymic stromal cells allow for the selection of a functional and self-tolerant T cell repertoire. Therefore, one of the most important roles of the thymus is the induction of central tolerance.

The thymus is largest and most active during the neonatal and pre-adolescent periods. By the early teens, the thymus begins to atrophy and thymic stroma is mostly replaced by adipose (fat) tissue. Nevertheless, residual T lymphopoiesis continues throughout adult life.

Structure

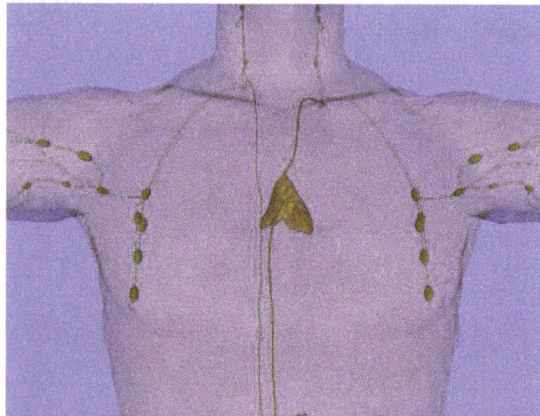

Anterior view of chest showing location and size of adult thymus

The thymus is of a pinkish-gray color, soft, and lobulated on its surfaces. At birth it is about 5 cm in length, 4 cm in breadth, and about 6 mm in thickness. The organ enlarges during childhood, and atrophies at puberty. Unlike the liver, kidney and heart, for instance, the thymus is at its largest in children. The thymus reaches maximum weight (20 to 37 grams) by the time of puberty. The thymus of older people is scarcely distinguishable from surrounding fatty tissue. As one ages the thymus slowly shrinks, eventually degenerating into tiny islands of fatty tissue. By the age of 75 years, the thymus weighs only 6 grams. In children the thymus is grayish-pink in colour and in adults it is yellow.

If examined when its growth is most active, the thymus will be found to consist of two lateral lobes placed in close contact along the middle line, situated partly in the thorax, partly in the neck, and extending from the fourth costal cartilage upward, as high as the lower border of the thyroid gland. It is covered by the sternum, and by the origins of the sternohyoid and sternothyroid muscles. Below, it rests upon the pericardium, being separated from the aortic arch and great vessels by a layer of fascia. In the neck, it lies on the front and sides of the trachea, behind the sternohyoidei and sternothyreoidei. The two lobes differ slightly in size and may be united or separated.

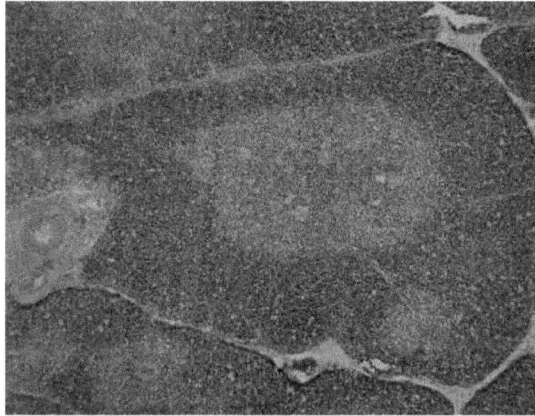

Histology

Each lateral lobe is composed of numerous lobules held together by delicate areolar tissue; the entire organ being enclosed in an investing capsule of a similar but denser structure. The primary lobules vary in size from that of a pin's head to that of a small pea, and are made up of a number of small nodules or follicles.

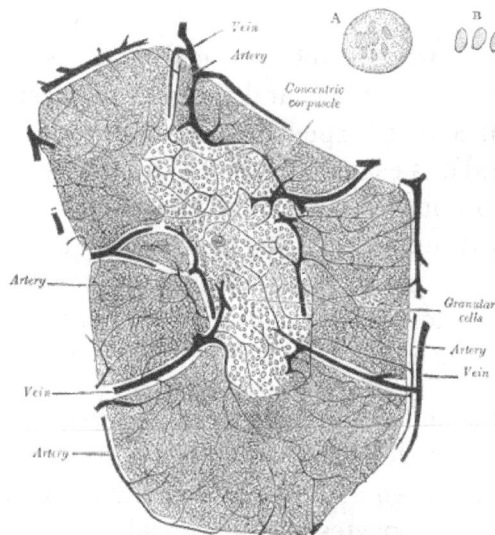

Minute structure of thymus.

The follicles are irregular in shape and are more or less fused together, especially toward the interior of the organ. Each follicle is from 1 to 2 mm in diameter and consists of a medullary and a cortical portion, and these differ in many essential particulars from each other.

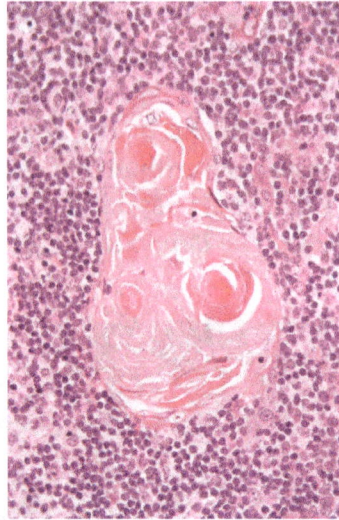

Micrograph showing a thymic corpuscle (Hassall corpuscle), a characteristic histologic feature of the human thymus.
H&E stain.

Cortex

The cortical portion is mainly composed of lymphocytes, supported by a network of finely-branched epithelial reticular cells, which is continuous with a similar network in the medullary portion. This network forms an adventitia to the blood vessels.

The cortex is the location of the earliest events in thymocyte development, where T cell receptor gene rearrangement and positive selection takes place.

Medulla

In the medullary portion, the network of reticular cells is coarser than in the cortex, the lymphoid cells are relatively fewer in number, and there are concentric, nest-like bodies called Hassall's corpuscles. These concentric corpuscles are composed of a central mass, consisting of one or more granular cells, and of a capsule formed of epithelial cells. They are the remains of the epithelial tubes, which grow out from the third pharyngeal pouches of the embryo to form the thymus. Each follicle is surrounded by a vascular plexus, from which vessels pass into the interior, and radiate from the periphery toward the center, forming a second zone just within the margin of the medullary portion. In the center of the medullary portion there are very few vessels, and they are of minute size.

The medulla is the location of the latter events in thymocyte development. Thymocytes that reach the medulla have already successfully undergone T cell receptor gene rearrangement and positive selection, and have been exposed to a limited degree of negative selection. The medulla is specialised to allow thymocytes to undergo additional rounds of negative selection to remove auto-reactive T-cells from the mature repertoire. The gene AIRE is expressed by the thymic medullary epithelium, and drives the transcription of organ-specific genes such as insulin to allow maturing thymocytes to be exposed to a more complex set of self-antigens than is present in the cortex.

Blood Supply

The arteries supplying the thymus are derived from the internal thoracic artery, and from the superior thyroid artery and inferior thyroids.

The veins end in the left brachiocephalic vein (innominate vein), and in the thyroid veins.

The nerves are exceedingly minute; they are derived from the vagi and sympathetic nervous system. Branches from the descendens hypoglossi and phrenic reach the investing capsule, but do not penetrate into the substance of the organ.

Development

The two main components of the thymus, the lymphoid thymocytes and the thymic epithelial cells, have distinct developmental origins. The thymic epithelium is the first to develop, and appears in the form of two flask-shape endodermal diverticula, which arise, one on either side, from the third pharyngeal pouch, and extend lateralward and backward into the surrounding mesoderm and neural crest-derived mesenchyme in front of the ventral aorta. Here the thymocytes and epithelium meet and join with connective tissue. The pharyngeal opening of each diverticulum is soon obliterated, but the neck of the flask persists for some time as a cellular cord. By further proliferation of the cells lining the flask, buds of cells are formed, which become surrounded and isolated by the invading mesoderm. Additional portions of thymus tissue are sometimes developed from the fourth pharyngeal pouch.

During the late stages of the development of the thymic epithelium, hematopoietic bone-marrow precursors migrate into the thymus. Normal thymic development thereafter is dependent on the interaction between the thymic epithelium and the hematopoietic thymocytes.

Thymic Involution

The thymus continues to grow between birth and puberty and then begins to atrophy; this thymic involution is directed by the high levels of circulating hormones. Proportional to thymic size, thymic activity (T-cell output) is most active before puberty. Upon atrophy, the size and activity are dramatically reduced, and the organ is primarily replaced with fat (a phenomenon known as "organ involution"). The atrophy is due to the increased circulating level of sex hormones, and chemical or physical castration of an adult results in the thymus increasing in size and activity. Patients with the autoimmune disease myasthenia gravis commonly (70%) are found to have thymic hyperplasia or malignancy. The reason or order of these circumstances has yet to be determined.

Age	Mass
birth	about 15 grams
puberty	about 35 grams
twenty-five years	25 grams
sixty years	less than 15 grams
seventy years	as low as 5 grams

Function

In the two thymic lobes, hematopoietic precursors from the bone-marrow, referred to as thymocytes, mature into T-cells. Once mature, T-cells emigrate from the thymus and constitute the peripheral T-cell repertoire[?] responsible for directing many facets of the adaptive immune system. Loss of the thymus at an early age through genetic mutation (as in DiGeorge Syndrome) results in severe immunodeficiency and subsequent high susceptibility to infection.

Each T cell attacks a specific foreign substance which it identifies with its receptor. T cells have receptors which are generated by randomly shuffling gene segments. Each T cell attacks a different antigen. T cells that attack the body's own proteins are eliminated in the thymus. Thymic epithelial cells express major proteins from elsewhere in the body. First, T cells undergo "Positive Selection", whereby the cell comes in contact with self-MHC, expressed by thymic epithelial cells; those with no interaction are destroyed. Second, the T cell undergoes "Negative Selection" by interacting with thymic dendritic cells, whereby T cells with high affinity interaction are eliminated through apoptosis (to avoid autoimmunity), and those with intermediate affinity survive.

The stock of T-lymphocytes is built up in early life, so the function of the thymus is diminished in adults. It is largely degenerated in elderly adults and is barely identifiable, consisting mostly of fatty tissue, but it continues its endocrine function. Involution of the thymus has been linked to loss of immune function in the elderly, susceptibility to infection and to cancer.

The ability of T cells to recognize foreign antigens is mediated by the T-cell receptor. The T-cell receptor undergoes genetic rearrangement during thymocyte maturation, resulting in each T-cell bearing a unique T-cell receptor, specific to a limited set of peptide:MHC combinations. The random nature of the genetic rearrangement results in a requirement of central tolerance mechanisms to remove or inactivate those T cells which bear a T-cell receptor with the ability to recognise self-peptides.

1. A rare population of hematopoietic progenitor cells enter the thymus from the blood, and expands by cell division to generate a large population of immature thymocytes.

2. Immature thymocytes each make distinct T-cell receptors by a process of gene rearrangement. This process is error-prone, and some thymocytes fail to make functional T-cell receptors, whereas other thymocytes make T-cell receptors that are autoreactive.

3. Immature thymocytes undergo a process of selection, based on the specificity of their T-cell receptors. This involves selection of T-cells that are *functional (positive selection)*, and elimination of T-cells that are *autoreactive (negative selection)*. The medulla of the thymus is the site of T Cell maturation.

type:	functional (positive selection)	autoreactive (negative selection)
location:	cortex	medulla

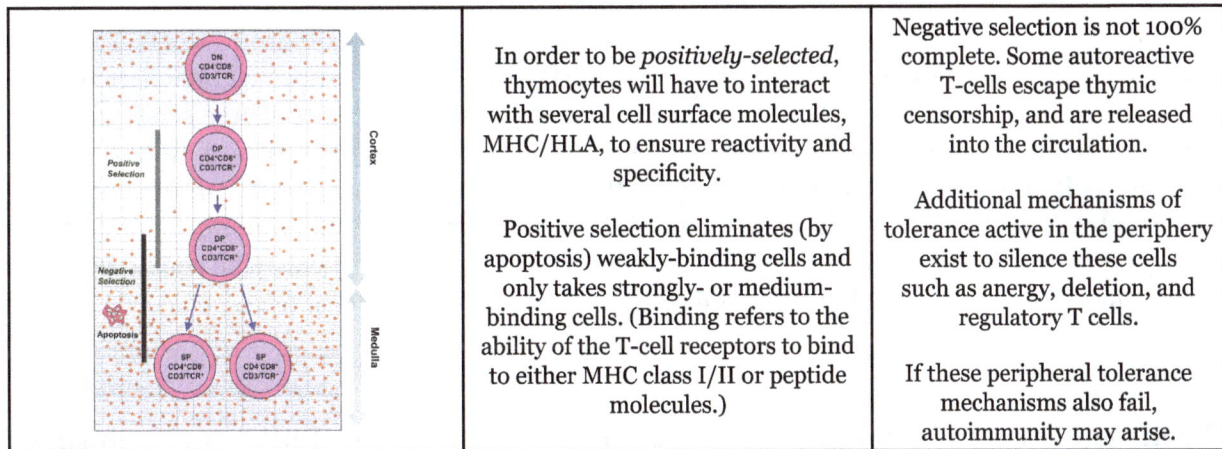

In order to be *positively-selected*, thymocytes will have to interact with several cell surface molecules, MHC/HLA, to ensure reactivity and specificity.

Positive selection eliminates (by apoptosis) weakly-binding cells and only takes strongly- or medium-binding cells. (Binding refers to the ability of the T-cell receptors to bind to either MHC class I/II or peptide molecules.)

Negative selection is not 100% complete. Some autoreactive T-cells escape thymic censorship, and are released into the circulation.

Additional mechanisms of tolerance active in the periphery exist to silence these cells such as anergy, deletion, and regulatory T cells.

If these peripheral tolerance mechanisms also fail, autoimmunity may arise.

Cells that pass both levels of selection are released into the bloodstream to perform vital immune functions.

Clinical Significance

The immune system is a multicomponent interactive system. It effectively protects the host from various infections. An improperly functioning immune system can cause discomfort, disease or even death. The type of malfunction falls into one or more of the following major groups: hypersensitivity or allergy, auto-immune disease, or immunodeficiency.

Hypersensitivity

Allergy results from an inappropriate and excessive immune response to common antigens. Substances that trigger an allergic response are called allergens. Allergies involve mainly IgE, antibodies, and histamine. Mast cells release the histamine. Sometimes an allergen may cause a sudden and severe, possibly fatal reaction in a sensitive individual; this is called anaphylaxis.

Immunodeficiency

As the thymus is the organ of T-cell development, any congenital defect in thymic genesis or a defect in thymocyte development can lead to a profound T cell deficiency in primary immunodeficiency disease. Defects that affect both the T cell and B cell lymphocyte lineages result in Severe Combined Immunodeficiency Syndrome (SCIDs). Acquired T cell deficiencies can also affect thymocyte development in the thymus.

Digeorge Syndrome

DiGeorge syndrome is a genetic disorder caused by the deletion of a small section of chromosome 22. This results in a midline congenital defect including thymic aplasia, or congenital deficiency of a thymus. Patients may present with a profound immunodeficiency disease, due to the lack of T cells. No other immune cell lineages are affected by the congenital absence of the thymus. DiGeorge syndrome is the most common congenital cause of thymic aplasia in humans. In mice, the nude mouse strain are congenitally thymic deficient. These mice are an important model of primary T cell deficiency.

SCID

Severe combined immunodeficiency syndromes (SCID) are group of rare congenital genetic diseases that result in combined T lymphocyte and B lymphocyte deficiencies. These syndromes are caused by defective hematopoietic progenitor cells which are the precursors of both B- and T-cells. This results in a severe reduction in developing thymocytes in the thymus and consequently thymic atrophy. A number of genetic defects can cause SCID, including IL-7 receptor deficiency, common gamma chain deficiency, and recombination activating gene deficiency. The gene that codes for the enzyme called ADA (adenine deaminase), is located on chromosomes 20.

HIV/AIDS

The HIV virus causes an acquired T-cell immunodeficiency syndrome (AIDS) by specifically killing CD4$^+$ T-cells. Whereas the major effect of the virus is on mature peripheral T-cells, HIV can also infect developing thymocytes in the thymus, most of which express CD4.

Autoimmune Disease

Autoimmune diseases are caused by a hyperactive immune system that instead of attacking foreign pathogens reacts against the host organism (self) causing disease. One of the primary functions of the thymus is to prevent autoimmunity through the process of central tolerance, immunologic tolerance to self antigens.

APECED

Autoimmune polyendocrinopathy-candidiasis-ectodermal dystrophy (APECED) is an extremely rare genetic autoimmune syndrome. However, this disease highlights the importance of the thymus in prevention of autoimmunity. This disease is caused by mutations in the Autoimmune Regulator (AIRE) gene. AIRE allows for the ectopic expression of tissue-specific proteins in the thymus medulla, such as proteins that would normally only be expressed in the eye or pancreas. This expression in the thymus, allows for the deletion of autoreactive thymocytes by exposing them to self-antigens during their development, a mechanism of central tolerance. Patients with APECED develop an autoimmune disease that affects multiple endocrine tissues.

Myasthenia Gravis

Myasthenia gravis is an autoimmune disease caused by antibodies that block acetylcholine receptors. Myasthenia gravis is often associated with thymic hypertrophy. Thymectomy may be necessary to treat the disease.

Cancer

Two primary forms of tumours originate in the thymus.

Thymomas

Tumours originating from the thymic epithelial cells are called thymomas, and are found in about 10-

15% of patients with myasthenia gravis. Symptoms are sometimes confused with bronchitis or a strong cough because the tumour presses on the recurrent laryngeal nerve. All thymomas are potentially cancerous, but they can vary a great deal. Some grow very slowly. Others grow rapidly and can spread to surrounding tissues. Treatment of thymomas often requires surgery to remove the entire thymus.

Lymphomas

Tumours originating from the thymocytes are called thymic lymphomas. Lymphomas or leukemias of thymocyte origin are classified as Precursor T acute lymphoblastic leukemia/lymphoma (T-ALL).

People with an enlarged thymus, particularly children, were treated with intense radiation in the years before 1950. There is an elevated incidence of thyroid cancer and leukemia in treated individuals.

Cervical Thymic Cyst

Cervical thymus is a rare malformation. Thymic tissue containing cysts is rarely described in the literature, ectopic glandular tissue included in the wall of cystic formation can trigger a series of problems similar to those of thymus.

Thymic cysts are uncommon lesions, about 150 cases being found. While thymic cyst and ectopic cervical thymus are identified most frequently in childhood, the mean age at which thymoma is diagnosed is 45 years. However, studies have shown the existence necroptic thymic tissue masses in the neck (asymptomatic intravital) more frequently, the incidence reaching nearly 30%. These observations may mean absence of clinical observation.

Thymectomy

Thymectomy is the surgical removal of the thymus. The usual reason for a thymectomy is to gain access to the heart for surgery to correct congenital heart defects in the neonatal period. In neonates, but not older children or adults, the relative size of the thymus obstructs surgical access to the heart. Removal of the thymus in infancy results in immunodeficiency by some measures, although T cells develop compensating function and it remains unknown whether disease incidence in later life is significantly greater. This is because sufficient T cells are generated during fetal life prior to birth. These T cells are long-lived and can proliferate by homeostatic proliferation throughout the lifetime of the patient. However, there is evidence of premature immune aging in patients thymectomized during early childhood.

Other indications for thymectomy include the removal of thymomas and the treatment of myasthenia gravis. Thymectomy is not indicated for the treatment of primary thymic lymphomas. However, a thymic biopsy may be necessary to make the pathologic diagnosis.

Therapeutical Approaches

Thymus Transplantation

A thymus may be transplanted, however, this approach is problematic due to donor requirements and matching tissue with the patient.

Thymus Tissue Engineering

A fully functional thymus derived from reprogrammed mouse embryonic fibroblasts has been grown in the kidney capsule of mice. The newly formed organ resembled a normal thymus histological and molecularly, and upon transplantation it was able to restore immune function in immunocompromised mice. The mouse embryonic fibroblasts were reprogrammed into thymic epithelial cells (TECs) by enforcing the expression of one transcription factor, *FOXN1*.

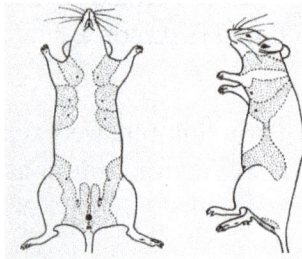

Society and Culture

When used for consumption, animal thymic tissue is known as (one of the kinds of) sweetbread.

History

Galen was the first to note that the size of the organ changed over the duration of a person's life.

In the nineteenth century, a fictitious condition known as *status thymicolymphaticus* (an "enlarged" thymus) was erroneously thought to be a cause of Sudden infant death syndrome, causing physicians to recommend radiation therapy as a treatment, a practice that continued into the 1950s.

Due to the large numbers of apoptotic lymphocytes, the thymus was originally dismissed as a "lymphocyte graveyard", without functional importance. The importance of the thymus in the immune system was discovered in 1961 by Jacques Miller, by surgically removing the thymus from one day old mice, and observing the subsequent deficiency in a lymphocyte population, subsequently named T-cells after the organ of their origin. Recently, advances in immunology have allowed the function of the thymus in T-cell maturation to be more fully understood.

Other Animals

The thymus is present in all jawed vertebrates, where it undergoes the same shrinkage with age and plays the same immunological function as in human beings. Recently, a discrete thymus-like lympho-epithelial structure, termed the thymoid, was discovered in the gills of larval lampreys. Hagfish possess a protothymus associated with the pharyngeal velar muscles, which is responsible for a variety of immune responses. Little is known about the immune mechanisms of tunicates or of *Amphioxus*.

The thymus is also present in most vertebrates, with similar structure and function as the human thymus. Some animals have multiple secondary (smaller) thymi in the neck; this phenomenon has been reported for mice and also occurs in 5 out of 6 human fetuses. As in humans, the Guinea pig's thymus naturally atrophies as the animal reaches adulthood, but the athymic hairless guinea pig (which arose from a spontaneous laboratory mutation) possesses no thymic tissue whatsoever, and the organ cavity is replaced with cystic spaces.

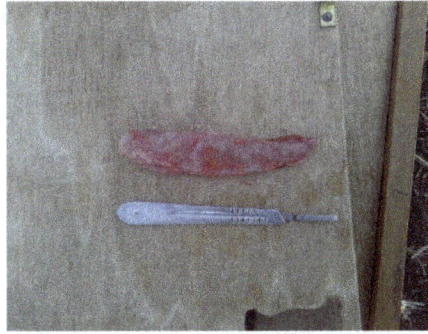

A sheep thymus, several times enlarged, in Peste des petits ruminants

Spleen

The spleen (from Greek σπλήν—*splēn*) is an organ found in virtually all vertebrates. Similar in structure to a large lymph node, it acts primarily as a blood filter.

The spleen plays important roles in regard to red blood cells (also referred to as erythrocytes) and the immune system. It removes old red blood cells and holds a reserve of blood, which can be valuable in case of hemorrhagic shock, and also recycles iron. As a part of the mononuclear phagocyte system, it metabolizes hemoglobin removed from senescent erythrocytes. The globin portion of hemoglobin is degraded to its constitutive amino acids, and the heme portion is metabolized to bilirubin, which is removed in the liver.

The spleen synthesizes antibodies in its white pulp and removes antibody-coated bacteria and antibody-coated blood cells by way of blood and lymph node circulation. A study published in 2009 using mice found that the red pulp of the spleen forms a reservoir that contains over half of the body's monocytes. These monocytes, upon moving to injured tissue (such as the heart after myocardial infarction), turn into dendritic cells and macrophages while promoting tissue healing. The spleen is a center of activity of the mononuclear phagocyte system and can be considered analogous to a large lymph node, as its absence causes a predisposition to certain infections.

In humans, the spleen is brownish in color and is located in the left upper quadrant of the abdomen.

Structure

The spleen, in healthy adult humans, is approximately 7 centimetres (2.8 in) to 14 centimetres (5.5 in) in length. It usually weighs between 150 grams (5.3 oz) and 200 grams (7.1 oz). An easy way to remember the anatomy of the spleen is the 1×3×5×7×9×11 rule. The spleen is 1" by 3" by 5", weighs approximately 7 oz, and lies between the 9th and 11th ribs on the left hand side.

Surfaces

The diaphragmatic surface of the spleen (or phrenic surface) is convex, smooth, and is directed upward, backward, and to the left, except at its upper end, where it is directed slightly to the middle.

It is in relation with the under surface of the diaphragm, which separates it from the ninth, tenth, and eleventh ribs of the left side, and the intervening lower border of the left lung and pleura.

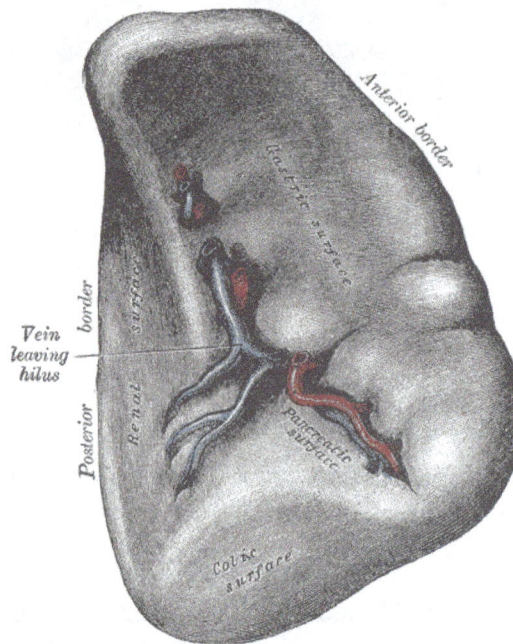

Visceral surface of the spleen

The visceral surface of the spleen is divided by a ridge into two regions: an anterior or gastric and a posterior or renal. The gastric surface (*facies gastrica*) is directed forward, upward, and toward the middle, is broad and concave, and is in contact with the posterior wall of the stomach. Below this it is in contact with the tail of the pancreas. Near to its mid-border is a long fissure, the splenic hilum. The hilum is the point of attachment for the gastrosplenic ligament, and the point of insertion for the splenic artery and splenic vein. There are other openings present for lymphatic vessels and nerves.

The renal (kidney) surface (*facies renalis*) is directed medialward and downward. It is somewhat flattened, considerably narrower than the gastric surface, and is in relation with the upper part of the anterior surface of the left kidney and occasionally with the left adrenal gland.

Like the thymus, the spleen possesses only efferent lymphatic vessels. The spleen is part of the lymphatic system. Both the short gastric arteries and the splenic artery supply it with blood.

The germinal centers are supplied by arterioles called *penicilliary radicles*.

Development

The spleen is unique in respect to its development within the gut. While most of the gut organs are endodermally derived (with the exception of the neural-crest derived adrenal gland), the spleen is derived from mesenchymal tissue. Specifically, the spleen forms within, and from, the dorsal mesentery. However, it still shares the same blood supply—the celiac trunk—as the foregut organs.

Function

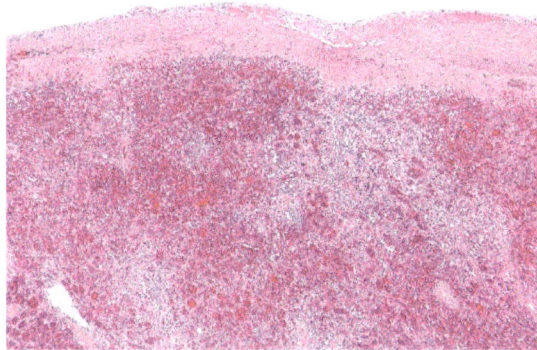

Micrograph of splenic tissue showing the red pulp (red), white pulp (blue) and a thickened inflamed capusule (mostly pink - top of image). H&E stain.

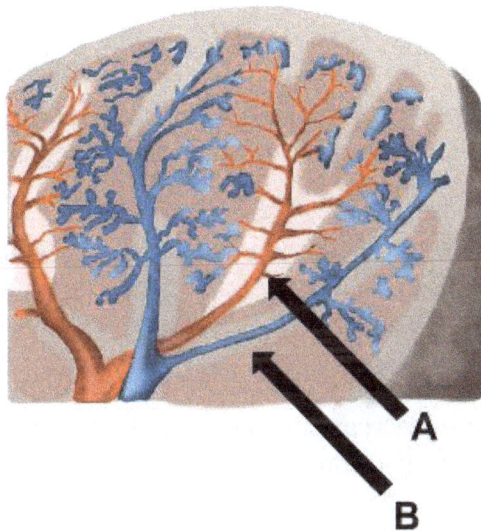

The spleen contains two different tissues, white pulp (A) and red pulp (B). The white pulp functions in producing and growing immune and blood cells. The red pulp functions in filtering blood of antigens, microorganisms, and defective or worn-out red blood cells.

Area	Function	Composition
red pulp	Mechanical filtration of red blood cells. In mice: Reserve of monocytes	• "sinuses" (or "sinusoids"), which are filled with blood • "splenic cords" of reticular fibers • "marginal zone" bordering on white pulp
white pulp	Active immune response through humoral and cell-mediated pathways.	Composed of nodules, called Malpighian corpuscles. These are composed of: • "lymphoid follicles" (or "follicles"), rich in B-lymphocytes • "periarteriolar lymphoid sheaths" (PALS), rich in T-lymphocytes

Other functions of the spleen are less prominent, especially in the healthy adult:

- Production of opsonins, properdin, and tuftsin.

- Creation of red blood cells. While the bone marrow is the primary site of hematopoiesis in the adult, the spleen has important hematopoietic functions up until the fifth month of gestation. After birth, erythropoietic functions cease, except in some hematologic disorders. As a major lymphoid organ and a central player in the reticuloendothelial system, the spleen retains the ability to produce lymphocytes and, as such, remains a hematopoietic organ.

- Storage of red blood cells, lymphocytes and other formed elements. In horses, roughly 30% of the red blood cells are stored there. The red blood cells can be released when needed. In humans, up to a cup (240 ml) of red blood cells can be held in the spleen and released in cases of hypovolemia. It can store platelets in case of an emergency and also clears old platelets from the circulation. Up to a quarter of lymphocytes can be stored in the spleen at any one time.

Clinical Significance

Enlarged Spleen

Disorders include splenomegaly, where the spleen is enlarged for various reasons, such as cancer, specifically blood-based leukemias, and asplenia, where the spleen is not present or functions abnormally.

Traumas, such as a road traffic collision, can cause rupture of the spleen, which is a situation requiring immediate medical attention.

Spleen Deflation

Human spleens have been noted to decrease in volume up to 40% when subjected to stimuli from strenuous exercise or hypoxic gas inhalation.

Decreased Function

Asplenia refers to a non-functioning spleen, which may be congenital, damaged by trauma, or caused by disease such as sickle cell anaemia. Hyposplenia refers to a partially functioning spleen. These conditions may cause: a modest increase in circulating white blood cells and platelets; a diminished response to some vaccines, and an increased susceptibility to infection. In particular, there is an increased risk of sepsis from polysaccharide encapsulated bacteria. Encapsulated bacteria inhibit binding of complement or prevent complement assembled on the capsule from interacting with macrophage receptors. Phagocytosis needs natural antibodies, which are immunoglobulins that facilitate phagocytosis either directly or by complement deposition on the capsule. They are produced by IgM memory B cells (a subtype of B cells) in the marginal zone of the spleen.

A splenectomy (removal of the spleen) also causes asplenia and results in a greatly diminished frequency of memory B cells. A 28-year follow-up of 740 World War II veterans whose spleens were removed on the battlefield showed a significant increase in the usual death rate from pneumonia (6 rather than the expected 1.3) and an increase in the death rate from ischemic heart disease (41 rather than the expected 30), but not from other conditions.

Accessory Spleen

An accessory spleen is a small splenic nodule extra to the spleen usually formed in early embryogenesis. Accessory spleens are found in approximately 10 percent of the population and are typically around 1 centimeter in diameter. *Splenosis* is a condition where displaced pieces of splenic tissue (often following trauma or splenectomy) autotransplant in the abdominal cavity as accessory spleens.

Polysplenia is a congenital disease manifested by multiple small accessory spleens, rather than a single, full-sized, normal spleen. Polysplenia sometimes occurs alone, but it is often accompanied by other developmental abnormalities such as intestinal malrotation or biliary atresia, or cardiac abnormalities, such as dextrocardia. These accessory spleens are non-functional.

Society and Culture

In English the word *spleen* was customary during the period of the 18th century. Authors like Richard Blackmore or George Cheyne employed it to characterise the hypochondriacal and hysterical affections. William Shakespeare, in *Julius Caesar* uses the spleen to describe Cassius' irritable nature.

Must I observe you? must I stand and crouch Under your testy humour? By the gods You shall digest the venom of your spleen, Though it do split you; for, from this day forth, I'll use you for my mirth, yea, for my laughter, When you are waspish.

In French, "splénétique" refers to a state of pensive sadness or melancholy. It has been popularized by the poet Charles Baudelaire (1821–1867) but was already used before in particular to the Romantic literature (19th century). The word for the organ is "rate".

The connection between *spleen* (the organ) and *melancholy* (the temperament) comes from the humoral medicine of the ancient Greeks. One of the humours (body fluid) was the black bile, secreted by the spleen organ and associated with melancholy. In contrast, the Talmud (tractate Berachoth 61b) refers to the spleen as the organ of laughter while possibly suggesting a link with the humoral view of the organ. In eighteenth- and nineteenth-century England, women in bad humor were said to be afflicted by the spleen, or the vapours of the spleen. In modern English, "to vent one's spleen" means to vent one's anger, e.g. by shouting, and can be applied to both males and females. Similarly, the English term "splenetic" is used to describe a person in a foul mood.

Other Animals

In cartilaginous and ray-finned fish, it consists primarily of red pulp and is normally somewhat elongated, as it lies inside the serosal lining of the intestine. In many amphibians, especially frogs, it has the more rounded form and there is often a greater quantity of white pulp.

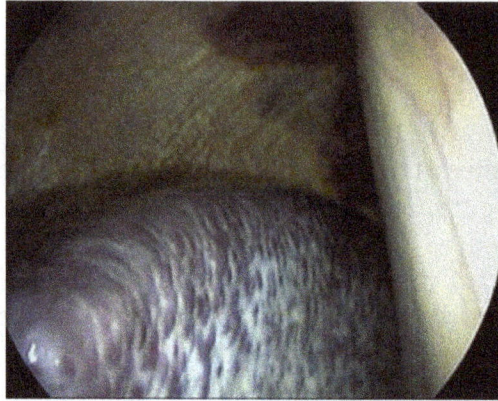
Laparoscopic view of a horse's spleen (the purple and grey mottled organ)

In reptiles, birds, and mammals, white pulp is always relatively plentiful, and in birds and mammals the spleen is typically rounded, but it adjusts its shape somewhat to the arrangement of the surrounding organs. In most vertebrates, the spleen continues to produce red blood cells throughout life; only in mammals this function is lost in adults. Many mammals have tiny spleen-like structures known as haemal nodes throughout the body that are presumed to have the same function as the spleen. The spleens of aquatic mammals differ in some ways from those of fully land-dwelling mammals; in general they are bluish in colour. In cetaceans and manatees they tend to be quite small, but in deep diving pinnipeds, they can be quite massive, due to their function of storing red blood cells.

The only vertebrates lacking a spleen are the lampreys and hagfishes (the Cyclostomata). Even in these animals, there is a diffuse layer of haematopoeitic tissue within the gut wall, which has a similar structure to red pulp and is presumed to be homologous with the spleen of higher vertebrates.

In mice, the spleen stores half the body's monocytes so that upon injury, they can migrate to the injured tissue and transform into dendritic cells and macrophages and so assist wound healing.

Lymph Node

A lymph node is an oval or kidney-shaped organ of the lymphatic system, present widely throughout the body including the armpit and stomach and linked by lymphatic vessels. Lymph nodes are major sites of B, T, and other immune cells. Lymph nodes are important for the proper functioning of the immune system, acting as filters for foreign particles and cancer cells. Lymph nodes do not deal with toxicity, which is primarily dealt with by the liver and kidneys.

Lymph nodes also have clinical significance. They become inflamed or enlarged in various infections and diseases which may range from trivial throat infections, to life-threatening cancers. The condition of the lymph nodes is very important in cancer staging, which decides the treatment to be used, and determines the prognosis. When swollen, inflamed or enlarged, lymph nodes can be hard, firm or tender.

Structure

Lymph nodes are kidney or oval shaped and range in size from a few millimeters to about 1–2 cm long. Each lymph node is surrounded by a fibrous capsule, and inside the lymph node the fibrous capsule extends to form trabeculae. The substance of the lymph node is divided into the outer cortex and the inner medulla. The cortex is continuous around the medulla except at the hilum, where the medulla comes in direct contact with the hilum.

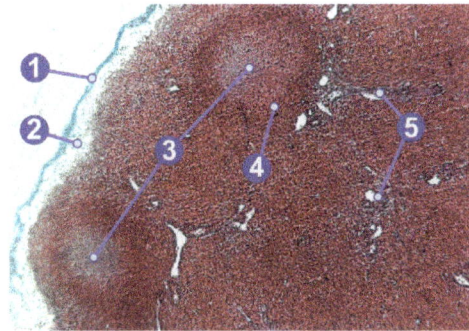

1) Capsule; 2) Subcapsular sinus; 3) Germinal centre; 4) Lymphoid nodule; 5) Trabeculae

Thin reticular fibers and elastin form a supporting meshwork called a *reticular network* inside the node. White blood cells (leukocytes), the most prominent ones being lymphocytes, are tightly packed in the follicles (B cells) and the cortex (T cells). Elsewhere in the node, there are only occasional leucocytes. As part of the reticular network there are follicular dendritic cells in the B cell follicle and fibroblastic reticular cells in the T cell cortex. The reticular network not only provides the structural support, but also the surface for adhesion of the dendritic cells, macrophages and lymphocytes. It allows exchange of material with blood through the high endothelial venules and provides the growth and regulatory factors necessary for activation and maturation of immune cells.

The number and composition of follicles can change especially when challenged by an antigen, when they develop a germinal center.

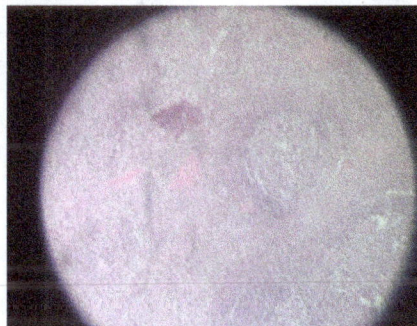

Lymph node histology

Lymph enters the convex side of the lymph node through multiple afferent lymphatic vessels, to flow through the sinuses. A lymph sinus which includes the subcapsular sinus, is a channel within the node, lined by endothelial cells along with fibroblastic reticular cells and this allows for the smooth flow of lymph through them.The endothelium of the subcapsular sinus is continuous with that of the afferent lymph vessel and is also with that of the similar sinuses flanking the trabeculae

and within the cortex. All of these sinuses drain the filtered lymphatic fluid into the medullary sinuses, from where the lymph flows into the efferent lymph vessels to exit the node at the hilum on
the concave side. These vessels are smaller and don't allow the passage of the macrophages so that
they remain contained to function within the lymph node. In the course of the lymph, lymphocytes
may be activated as part of the adaptive immune response.

Capsule

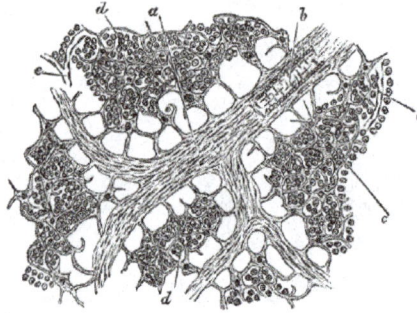

Lymph node tissue showing trabeculae

The lymph node capsule is composed of dense irregular connective tissue with some plain muscle
fibers, and from its internal surface are given off a number of membranous processes or trabeculae, consisting, in humans, of connective tissue, with a small admixture of plain muscle fibers; but
in many of the lower animals composed almost entirely of involuntary muscle. They pass inward,
radiating toward the center of the gland, for about one-third or one-fourth of the space between
the circumference and the center of the node. In some animals they are sufficiently well-marked
to divide the peripheral or cortical portion of the gland into a number of compartments (follicles),
but in humans this arrangement is not obvious. The larger trabeculae springing from the capsule
break up into finer bands, and these interlace to form a mesh-work in the central or medullary
portion of the gland. In these spaces formed by the interlacing trabeculae is contained the proper
gland substance or lymphoid tissue. The gland pulp does not, however, completely fill the spaces,
but leaves, between its outer margin and the enclosing trabeculae, a channel or space of uniform
width throughout. This is termed the subcapsular sinus (lymph path or lymph sinus). Running
across it are a number of finer trabeculæ of reticular connective tissue, the fibers of which are, for
the most part, covered by ramifying cells.

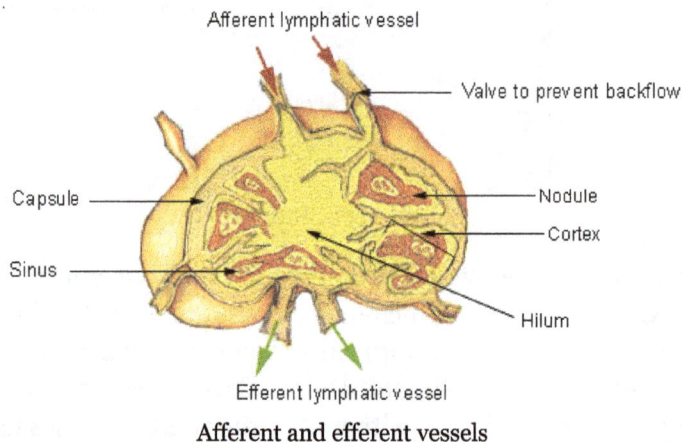

Afferent lymphatic vessel

Valve to prevent backflow

Capsule

Nodule

Cortex

Sinus

Hilum

Efferent lymphatic vessel

Afferent and efferent vessels

Subcapsular Sinus

The subcapsular sinus (lymph path, lymph sinus, marginal sinus) is the space between the capsule and the cortex which allows the free movement of lymphatic fluid and so contains a sparsity of lymphocytes. It is continuous with the similar lymph sinuses that flank the trabeculae.

The lymph node contains lymphoid tissue, i.e., a meshwork or fibers called *reticulum* with white blood cells enmeshed in it. The regions where there are few cells within the meshwork are known as *lymph sinus*. It is lined by reticular cells, fibroblasts and fixed macrophages.

The subcapsular sinus has clinical importance as it is the most likely location where the earliest manifestations of a metastatic carcinoma in a lymph node would be found.

Cortex

The cortex of the lymph node is the peripheral portion underneath the capsule and the subcapsular sinus. The subcapsular sinus drains to trabecular sinuses, and then the lymph flows into the medullary sinuses.

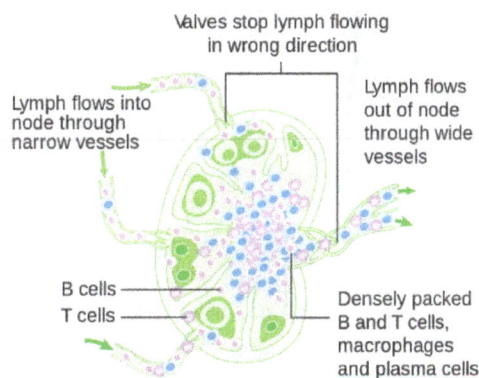

Diagram of a lymph node

The outer cortex consists mainly of the B cells arranged as follicles, which may develop a germinal center when challenged with an antigen, and the deeper cortex mainly consisting of the T cells. There is a zone known as the subcortical zone where T-cells (or cells that are mainly red) mainly interact with dendritic cells, and where the reticular network is dense. The predominant site within the lymph nodes which contain T cells & accessory cells is also known as paracortex (reticular network).

Medulla

The medulla contains large blood vessels, sinuses and medullary cords that contain antibody-secreting plasma cells.

The medullary cords are cords of lymphatic tissue, and include plasma cells, macrophages, and B cells. The medullary sinuses (or *sinusoids*) are vessel-like spaces separating the medullary cords. Lymph flows into the medullary sinuses from cortical sinuses, and into efferent lymphatic vessels. Medullary sinuses contain histiocytes (immobile macrophages) and reticular cells.

Function

The lymph fluid inside the lymph nodes contains lymphocytes, a type of white blood cell, which continuously recirculates through the lymph nodes and the bloodstream. Molecules found on bacteria cell walls or chemical substances secreted from bacteria, called antigens, may be taken up by dedicated antigen presenting cells such as dendritic cells into the lymph system and then into lymph nodes. In response to the antigens, the lymphocytes in the lymph node make antibodies which will go out of the lymph node into circulation, seek, and target the pathogens producing the antigens by targeting them for destruction by other cells. If the Lymphocytes in the lymph nodes decides the pathogen is too severe for it to handle alone it will activate the general immune system outside of the node for extra support. The increased numbers of immune system cells fighting the infection will make the node expand and become "swollen."

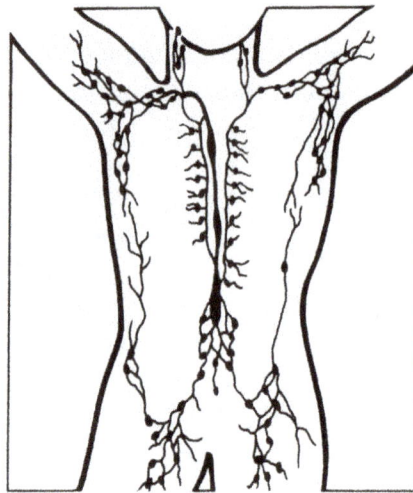

There are clusters of nodes under the arms, in the groin, neck and abdomen

Lymph circulates to the lymph node via *afferent lymphatic vessels* and drains into the node just beneath the capsule in a space called the subcapsular sinus. The subcapsular sinus drains into trabecular sinuses and finally into medullary sinuses. The sinus space is criss-crossed by the pseudopods of macrophages, which act to trap foreign particles and filter the lymph. The medullary sinuses converge at the hilum and lymph then leaves the lymph node via the *efferent lymphatic vessel* towards either a more central lymph node or ultimately for drainage into a central venous subclavian blood vessel.

- The B cells migrate to the nodular cortex and medulla.

- The T cells migrate to the deep cortex. This is a region of the lymph node called the paracortex that immediately surrounds the medulla. Unlike the cortex, which has mostly immature T cells, or thymocytes, the paracortex has a mixture of immature and mature T cells. Lymphocytes enter the lymph nodes through specialized high endothelial venules found in the paracortex.

When a lymphocyte recognizes an antigen, B cells become activated and migrate to germinal centers (by definition, a "secondary nodule" has a germinal center, while a "primary nodule" does not). When antibody-producing plasma cells are formed, they migrate to the medullary cords. Stimulation of the lymphocytes by antigens can accelerate the migration process to about 10 times normal, resulting in characteristic swelling of the lymph nodes.

The spleen and tonsils are large lymphoid organs that serve similar functions to lymph nodes, though the spleen filters blood cells rather than lymph.

Clinical Significance

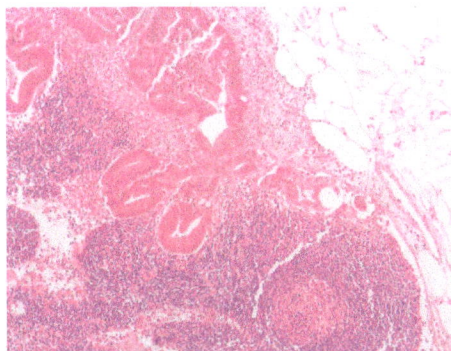

Micrograph of a mesenteric lymph node with adenocarcinoma

Lymph nodes may become enlarged due to an infection, tumour or inflamed due to leukemia. This increase in size is primarily due to an elevated rate of trafficking of lymphocytes into the node from the blood, exceeding the rate of outflow from the node. They may also be enlarged secondarily as a result of the activation and proliferation of antigen-specific T and B cells (clonal expansion). In some cases, where there is no infection they may still feel enlarged due to a previous infection. Lympho-granuloma venereum is a sexually transmitted infection that travels through the lymphatics and targets lymph nodes where the bacteria multiply. Enlarged and painful lymph nodes can result.

Lymphadenopathy is a term meaning "disease of the lymph nodes." It is, however, almost synonymously used with "swollen or enlarged lymph nodes." In this case, the lymph nodes are detectable by touch (palpable); this is a sign of various infections and diseases.

Lymphedema is a fairly widespread condition of the lymphatic system resulting in localised fluid retention and tissue swelling. Affected tissues are at risk of infection. Primary lymphedema generally results from poorly developed or missing lymph nodes. Secondary lymphedema results mostly from the dissection of lymph nodes during breast cancer surgery and also from other treatments usually involving radiation.

Lymphomas are classed as tumors of the hematopoietic and lymphoid tissues and mostly refer to malignancies. Lymph nodes may become swollen but are not often painful.

Bone Marrow

Bone marrow is the flexible tissue in the interior of bones. In humans, red blood cells are produced by cores of bone marrow in the heads of long bones in a process known as hematopoiesis. On average, bone marrow constitutes 4% of the total body mass of humans; in an adult having 65 kilograms of mass (143 lbs), bone marrow typically accounts for approximately 2.6 kilograms (5.7 lb). The hematopoietic component of bone marrow produces approximately 500 billion blood cells per day, which use the bone marrow vasculature as a conduit to the body's systemic circulation. Bone marrow is also a key component of the lymphatic system, producing the lymphocytes that support the body's immune system.

Bone marrow transplants can be conducted to treat severe diseases of the bone marrow, including certain forms of cancer such as leukemia. Additionally, bone marrow stem cells have been successfully transformed into functional neural cells, and can also potentially be used to treat illnesses such as inflammatory bowel disease.

Structure

Types of Bone Marrow

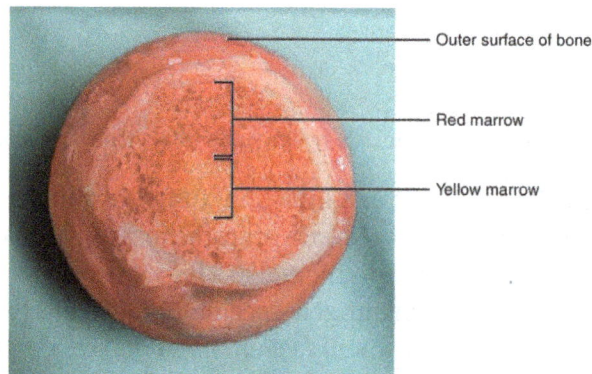

A femoral head with a cortex of bone and medulla of trabecular bone. Both red bone marrow and a central focus of yellow bone marrow are visible.

The two types of bone marrow are "red marrow" (Latin: *medulla ossium rubra*), which consists mainly of hematopoietic tissue, and "yellow marrow" (Latin: *medulla ossium flava*), which is mainly made up of fat cells. Red blood cells, platelets, and most white blood cells arise in red marrow. Both types of bone marrow contain numerous blood vessels and capillaries. At birth, all bone marrow is red. With age, more and more of it is converted to the yellow type; only around half of adult bone marrow is red. Red marrow is found mainly in the flat bones, such as the pelvis, sternum, cranium, ribs, vertebrae and scapulae, and in the cancellous ("spongy") material at the epiphyseal ends of long bones such as the femur and humerus. Yellow marrow is found in the medullary cavity, the hollow interior of the middle portion of short bones. In cases of severe blood loss, the body can convert yellow marrow back to red marrow to increase blood cell production.

Stroma

The stroma of the bone marrow is all tissue not directly involved in the marrow's primary function of hematopoiesis. Yellow bone marrow makes up the majority of bone marrow stroma, in addition to smaller concentrations of stromal cells located in the red bone marrow. Though not as active as parenchymal red marrow, stroma is indirectly involved in hematopoiesis, since it provides the hematopoietic microenvironment that facilitates hematopoiesis by the parenchymal cells. For instance, they generate colony stimulating factors, which have a significant effect on hematopoiesis.

Cell types that constitute the bone marrow stroma include:

- fibroblasts (reticular connective tissue)

- macrophages, which contribute especially to red blood cell production, as they deliver iron for hemoglobin production.

- adipocytes (fat cells)

- osteoblasts (synthesize bone)

- osteoclasts (resorb bone)

- endothelial cells, which form the sinusoids. These derive from endothelial stem cells, which are also present in the bone marrow.

Cellular Components

Cellular constitution of the red bone marrow parenchyma			
Group	**Cell type**	**Average fraction**	**Reference range**
Myelopoietic cells	Myeloblasts	0.9%	0.2-1.5
	Promyelocytes	3.3%	2.1-4.1
	Neutrophilic myelocytes	12.7%	8.2-15.7
	Eosinophilic myelocytes	0.8%	0.2-1.3
	Neutrophilic metamyelocytes	15.9%	9.6-24.6
	Eosinophilic metamyelocytes	1.2%	0.4-2.2
	Neutrophilic band cells	12.4%	9.5-15.3
	Eosinophilic band cells	0.9%	0.2-2.4
	Segmented neutrophils	7.4%	6.0-12.0
	Segmented eosinophils	0.5%	0.0-1.3
	Segmented basophils and mast cells	0.1%	0.0-0.2
Erythropoietic cells	Pronormoblasts	0.6%	0.2-1.3
	Basophilic normoblasts	1.4%	0.5-2.4
	Polychromatic normoblasts	21.6%	17.9-29.2
	Orthochromatic normoblast	2.0%	0.4-4.6
Other cell types	Megakaryocytes	< 0.1%	0.0-0.4
	Plasma cells	1.3%	0.4-3.9
	Reticular cells	0.3%	0.0-0.9
	Lymphocytes	16.2%	11.1-23.2
	Monocytes	0.3%	0.0-0.8

In addition, the bone marrow contains hematopoietic stem cells, which give rise to the three classes of blood cells that are found in the circulation: white blood cells (leukocytes), red blood cells (erythrocytes), and platelets (thrombocytes).

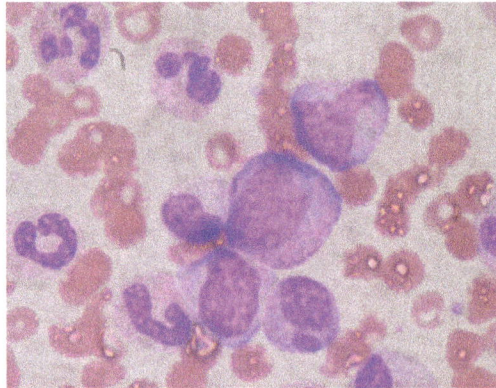

Hematopoietic precursor cells: promyelocyte in the center, two metamyelocytes next to it and band cells from a bone marrow aspirate.

Function

Mesenchymal Stem Cells

The bone marrow stroma contains mesenchymal stem cells (MSCs), also known as marrow stromal cells. These are multipotent stem cells that can differentiate into a variety of cell types. MSCs have been shown to differentiate, in vitro or in vivo, into osteoblasts, chondrocytes, myocytes, adipocytes and beta-pancreatic islets cells.

Bone Marrow Barrier

The blood vessels of the bone marrow constitute a barrier, inhibiting immature blood cells from leaving the marrow. Only mature blood cells contain the membrane proteins, such as aquaporin and glycophorin, that are required to attach to and pass the blood vessel endothelium. Hematopoietic stem cells may also cross the bone marrow barrier, and may thus be harvested from blood.

Lymphatic Role

The red bone marrow is a key element of the lymphatic system, being one of the primary lymphoid organs that generate lymphocytes from immature hematopoietic progenitor cells. The bone marrow and thymus constitute the primary lymphoid tissues involved in the production and early selection of lymphocytes. Furthermore, bone marrow performs a valve-like function to prevent the backflow of lymphatic fluid in the lymphatic system.

Compartmentalization

Biological compartmentalization is evident within the bone marrow, in that certain cell types tend to aggregate in specific areas. For instance, erythrocytes, macrophages, and their precursors tend to gather around blood vessels, while granulocytes gather at the borders of the bone marrow.

Society and Culture

Animal bone marrow has been used in cuisine worldwide for millennia, such as the famed Milanese Ossobuco.

Clinical Significance

Disease

The normal bone marrow architecture can be damaged or displaced by aplastic anemia, malignancies such as multiple myeloma, or infections such as tuberculosis, leading to a decrease in the production of blood cells and blood platelets. The bone marrow can also be affected by various forms of leukemia, which attacks its hematologic progenitor cells. Furthermore, exposure to radiation or chemotherapy will kill many of the rapidly dividing cells of the bone marrow, and will therefore result in a depressed immune system. Many of the symptoms of radiation poisoning are due to damage sustained by the bone marrow cells.

To diagnose diseases involving the bone marrow, a bone marrow aspiration is sometimes performed. This typically involves using a hollow needle to acquire a sample of red bone marrow from the crest of the ilium under general or local anesthesia.

Imaging

On CT and plain film, marrow change can be seen indirectly by assessing change to the adjacent ossified bone. Assessment with MRI is usually more sensitive and specific for pathology, particularly for hematologic malignancies like leukemia and lymphoma. These are difficult to distinguish from the red marrow hyperplasia of hematopoiesis, as can occur with tobacco smoking, chronically anemic disease states like sickle cell anemia or beta thalassemia, medications such as granulocyte colony-stimulating factors, or during recovery from chronic nutritional anemias or therapeutic bone marrow suppression. On MRI, the marrow signal is not supposed to be brighter than the adjacent intervertebral disc on T1 weighted images, either in the coronal or sagittal plane, where they can be assessed immediately adjacent to one another. Fatty marrow change, the inverse of red marrow hyperplasia, can occur with normal aging, though it can also be seen with certain treatments such as radiation therapy. Diffuse marrow T1 hypointensity without contrast enhancement or cortical discontinuity suggests red marrow conversion or myelofibrosis. Falsely normal marrow on T1 can be seen with diffuse multiple myeloma or leukemic infiltration when the water to fat ratio is not sufficiently altered, as may be seen with lower grade tumors or earlier in the disease process.

Histology

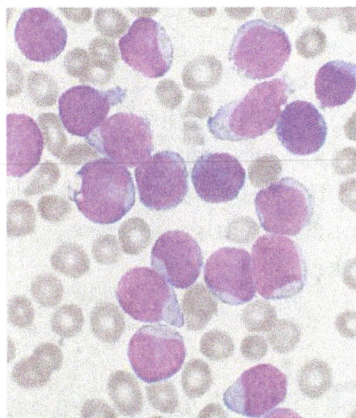

A Wright's-stained bone marrow aspirate smear from a patient with leukemia.

Bone marrow examination is the pathologic analysis of samples of bone marrow obtained via biopsy and bone marrow aspiration. Bone marrow examination is used in the diagnosis of a number of conditions, including leukemia, multiple myeloma, anemia, and pancytopenia. The bone marrow produces the cellular elements of the blood, including platelets, red blood cells and white blood cells. While much information can be gleaned by testing the blood itself (drawn from a vein by phlebotomy), it is sometimes necessary to examine the source of the blood cells in the bone marrow to obtain more information on hematopoiesis; this is the role of bone marrow aspiration and biopsy.

The ratio between myeloid series and erythroid cells is relevant to bone marrow function, and also to diseases of the bone marrow and peripheral blood, such as leukemia and anemia. The normal myeloid-to-erythroid ratio is around 3:1; this ratio may increase in myelogenous leukemias, decrease in polycythemias, and reverse in cases of thalassemia.

Donation and Transplantation

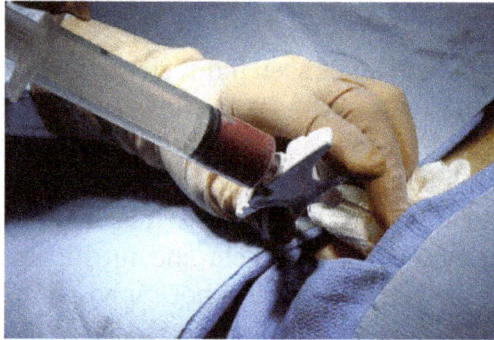

A bone marrow harvest in progress.

In a bone marrow transplant, hematopoietic stem cells are removed from a person and infused into another person (allogenic) or into the same person at a later time (autologous). If the donor and recipient are compatible, these infused cells will then travel to the bone marrow and initiate blood cell production. Transplantation from one person to another is conducted for the treatment of severe bone marrow diseases, such as congenital defects, autoimmune diseases or malignancies. The patient's own marrow is first killed off with drugs or radiation, and then the new stem cells are introduced. Before radiation therapy or chemotherapy in cases of cancer, some of the patient's hematopoietic stem cells are sometimes harvested and later infused back when the therapy is finished to restore the immune system.

Bone marrow stem cells can be induced to become neural cells to treat neurological illnesses, and can also potentially be used for the treatment of other illnesses, such as inflammatory bowel disease. In 2013, following a clinical trial, scientists proposed that bone marrow transplantation could be used to treat HIV in conjunction with antiretroviral drugs; however, it was later found that HIV remained in the bodies of the test subjects.

Harvesting

The stem cells are typically harvested directly from the red marrow in the iliac crest, often under general anesthesia. The procedure is minimally invasive and does not require stitches afterwards.

Depending on the donor's health and reaction to the procedure, the actual harvesting can be an outpatient procedure, or can require 1–2 days of recovery in the hospital.

Another option is to administer certain drugs that stimulate the release of stem cells from the bone marrow into circulating blood. An intravenous catheter is inserted into the donor's arm, and the stem cells are then filtered out of the blood. This procedure is similar to that used in blood or platelet donation. In adults, bone marrow may also be taken from the sternum, while the tibia is often used when taking samples from infants. In newborns, stem cells may be retrieved from the umbilical cord.

Fossil Record

Bone marrow may have first evolved in *Eusthenopteron*, a species of prehistoric fish with close links to early tetrapods.

The earliest fossilised evidence of bone marrow was discovered in 2014 in *Eusthenopteron*, a lobe-finned fish which lived during the Devonian period approximately 370 million years ago. Scientists from Uppsala University and the European Synchrotron Radiation Facility used X-ray synchrotron microtomography to study the fossilised interior of the skeleton's humerus, finding organised tubular structures akin to modern vertebrate bone marrow. *Eusthenopteron* is closely related to the early tetrapods, which ultimately evolved into the land-dwelling mammals and lizards of the present day.

Tonsil

Tonsils are collections of lymphoid tissue facing into the aerodigestive tract. The set of lymphatic tissue known as Waldeyer's tonsillar ring includes the adenoid tonsil, two tubal tonsils, two palatine tonsils, and the lingual tonsil.

When used unqualified, the term most commonly refers specifically to the palatine tonsils, which are masses of lymphatic material situated at either side at the back of the human throat. The palatine tonsils and the nasopharyngeal tonsil are lymphoepithelial tissues located near the oropharynx and nasopharynx (parts of the throat).

Structure

Tonsils in humans include, from anterior (front), superior (top), posterior (back), and inferior (bottom):

Type	Epithelium	capsule	Crypts	Location
Adenoids (also termed "pharyngeal tonsils")	Ciliated pseudostratified columnar (respiratory epithelium)	Incompletely encapsulated	No crypts, but small folds	Roof of pharynx
Tubal tonsils	Ciliated pseudostratified columnar (respiratory epithelium)			Roof of pharynx
Palatine tonsils	Non-keratinized stratified squamous	Incompletely encapsulated	Long, branched	Sides of oropharynx between palatoglossal and palatopharyngeal arches
Lingual tonsils	Non-keratinized stratified squamous	Incompletely encapsulated	Long, branched	Behind terminal sulcus (tongue)

Development

Tonsils tend to reach their largest size near puberty, and they gradually undergo atrophy thereafter. However, they are largest relative to the diameter of the throat in young children.

Function

These immunocompetent tissues are the immune system's first line of defense against ingested or inhaled foreign pathogens. Tonsils have on their surface specialized antigen capture cells called M cells that allow for the uptake of antigens produced by pathogens. These M cells then alert the underlying B cells and T cells in the tonsil that a pathogen is present and an immune response is stimulated. B cells are activated and proliferate in areas called germinal centres in the tonsil. These germinal centres are places where B memory cells are created and secretory antibody (IgA) is produced.

Recent studies have provided evidence that the tonsils produce T lymphocytes, also known as T-cells, in a manner similar to, but different from, the way the thymus does.

Clinical Significance

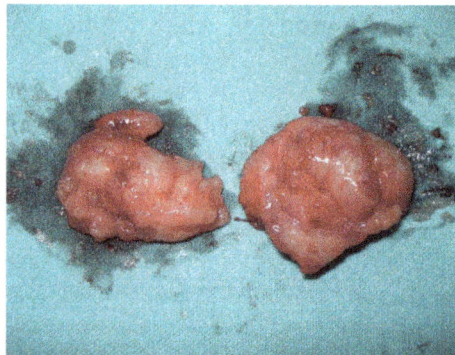

A pair of tonsils after surgical removal.

Tonsils can become enlarged (adenotonsillar hyperplasia) or inflamed (tonsillitis) and may require surgical removal (tonsillectomy). This may be indicated if they obstruct the airway or interfere with swallowing, or in patients with frequent recurrent tonsillitis. However, different mechanisms of pathogenesis for these two subtypes of tonsillar hypertrophy have been described, and may have different responses to identical therapeutic efforts. In older patients, asymmetric tonsils (also known as asymmetric tonsil hypertrophy) may be an indicator of virally infected tonsils, or tumors such as lymphoma or squamous cell carcinoma.

Tonsillitis is a disorder in which the tonsils are inflamed (sore and swollen). The most common way to treat it is with anti-inflammatory drugs such as ibuprofen, or if bacterial in origin, antibiotics, e.g. amoxicillin and azithromycin. Often severe and/or recurrent tonsillitis is treated by tonsillectomy.

A tonsillolith is material that accumulates on the tonsil. They can range up to the size of a peppercorn and are white/cream in color. The main substance is mostly calcium, but they have a strong unpleasant odor because of hydrogen sulfide and methyl mercaptan and other chemicals.

Tonsil enlargement can affect speech, making it hypernasal and giving it the sound of velopharyngeal incompetence (when space in the mouth is not fully separated from the nose's air space). Tonsil size may have a more significant impact on upper airway obstruction for obese children than for those of average weight.

As mucosal lymphatic tissue of the aerodigestive tract, the tonsils are viewed in some classifications as belonging to both the gut-associated lymphoid tissue (GALT) and the mucosa-associated lymphoid tissue (MALT). Other viewpoints treat them (and the spleen and thymus) as large lymphatic organs contradistinguished from the smaller tissue loci of GALT and MALT.

Additional Images

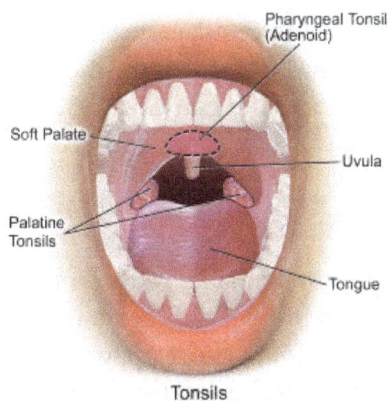

Illustration of frontal view of tonsils.

Adenoid

The adenoid, also known as a pharyngeal tonsil or nasopharyngeal tonsil, is the superior-most of the tonsils. It is a mass of lymphatic tissue situated posterior to the nasal cavity, in the roof of the nasopharynx, where the nose blends into the throat. Normally, in children, it forms a soft mound in the roof and posterior wall of the nasopharynx, just above and behind the uvula.

Structure

The adenoid, unlike the palatine tonsils, has pseudostratified epithelium. The adenoid is often removed along with the palatine tonsils.

Development

Adenoids develop from a subepithelial infiltration of lymphocytes after the 16th week of embryonic life. They are part of the so-called Waldeyer ring of lymphoid tissue which includes the palatine tonsils and the lingual tonsils.

After birth, enlargement begins and continues until aged 5 to 7 years. Symptomatic enlargement between 18 and 24 months of age is not uncommon, meaning that snoring, nasal airway obstruction and obstructed breathing may occur during sleep. However, this may be reasonably expected to decline when children reach school age, and progressive shrinkage may be expected thereafter.

The establishment of the upper respiratory tract is initiated at birth. Species of bacteria such as lactobacilli, anaerobic streptococci, actinomycosis, Fusobacterium species, and Nocardia are normally present by 6 months of age. Normal flora found in the adenoid consists of alpha-hemolytic streptococci and enterococci, Corynebacterium species, coagulase-negative staphylococci, Neisseria species, Haemophilus species, Micrococcus species, and Stomatococcus species.

Clinical Significance

An enlarged adenoid, or adenoid hypertrophy, can become nearly the size of a ping pong ball and completely block airflow through the nasal passages. Even if the enlarged adenoid is not substantial enough to physically block the back of the nose, it can obstruct airflow enough so that breathing through the nose requires an uncomfortable amount of work, and inhalation occurs instead through an open mouth. The enlarged adenoid can also obstruct the nasal airway enough to affect the voice without actually stopping nasal airflow altogether.

Adenoid Facies

Enlargement of the adenoid, especially in children, causes an atypical appearance of the face, often referred to as *adenoid facies*. Features of adenoid facies include mouth breathing, an elongated face, prominent incisors, hypoplastic maxilla, short upper lip, elevated nostrils, and a high arched palate. George Catlin, in his humorous and instructive book *Breath of Life*, published in 1861, illustrates adenoid facies in many engravings and advocates nose-breathing.

Removal

Surgical removal of the adenoid is a procedure called adenoidectomy. Adenoid infection may cause symptoms such as excessive mucus production, which can be treated by its removal. Studies have shown that adenoid regrowth occurs in as many as 20% of the cases after removal. Carried out through the mouth under a general anaesthetic (or less commonly a topical), adenoidectomy involves the adenoid being curetted, cauterized, lasered, or otherwise ablated.

References

- Tak W. Mak; Mary E. Saunders (Ph.D.); Mary E. Saunders (2008). Primer to the immune response. Academic Press. pp. 28–. ISBN 978-0-12-374163-9. Retrieved 12 November 2010.

- Anthony S. Fauci; Eugene Braunwald; Dennis Kasper; Stephen Hauser; Dan L. Longo (19 March 2009). Harrison's Manual of Medicine. McGraw Hill Professional. pp. 352–. ISBN 978-0-07-147743-7. Retrieved 12 November 2010.

- Britton, the editors Nicki R. Colledge, Brian R. Walker, Stuart H. Ralston ; illustated by Robert (2010). Davidson's principles and practice of medicine. (21st ed.). Edinburgh: Churchill Livingstone/Elsevier. pp. 1001, 1037–1040. ISBN 978-0-7020-3085-7.

- Histology image:07403loa from Vaughan, Deborah (2002). A Learning System in Histology: CD-ROM and Guide. Oxford University Press. ISBN 978-0195151732.

- Huete-Garin, A.; S.S. Sagel (2005). "Chapter 6: "Mediastinum", Thymic Neoplasm". In J.K.T. Lee; S.S. Sagel; R.J. Stanley; J.P. Heiken. Computed Body Tomography with MRI Correlation. Philadelphia: Lippincott Williams & Wilkins. pp. 311–324. ISBN 0-7817-4526-8.

- Sapolsky, Robert M. (2004). Why zebras don't get ulcers (3rd ed.). New York: Henry Hold and Co./Owl Books. pp. 182–185. ISBN 0805073698.

- Loscalzo, Joseph; Fauci, Anthony S.; Braunwald, Eugene; Dennis L. Kasper; Hauser, Stephen L; Longo, Dan L. (2008). Harrison's principles of internal medicine. McGraw-Hill Medical. ISBN 978-0-07-146633-2.

- Romer, Alfred Sherwood; Parsons, Thomas S. (1977). The Vertebrate Body. Philadelphia: Holt-Saunders International. pp. 410–411. ISBN 0-03-910284-X.

- Raphael Rubin & David S. Strayer (2007). Rubin's Pathology: Clinicopathologic Foundations of Medicine. Lippincott Williams & Wilkins. p. 90. ISBN 0-7817-9516-8.

- Bazigou E, Wilson J, Moore JE Primary and secondary lymphatic valve development: Molecular, functional and mechanical insights.Microvasc Res. 2014 Jul 30. pii: S0026-2862(14)00112-5. PMID 25086182

- "Research Supports Promise of Cell Therapy for Bowel Disease". Wake Forest Baptist Medical Center. 28 February 2013. Retrieved 5 March 2013.

- "HIV Returns in Two Men Thought 'Cured' by Bone Marrow Transplants". RH Reality Check. 10 December 2013. Retrieved 10 December 2013.

- Kato A et al, B-lymphocyte lineage cells and the respiratory system, Journal of Allergy and Clinical Immunology, Volume 131, pages 933-957, 2013

- "Tonsils Make T-Cells, Too, Ohio State Study Shows". Ohio State University. Ohio State University, Comprehensive Cancer Center. March 4, 2012. Retrieved March 27, 2014.

- "Circulating phospholipase-A2 activity in obstructive sleep apnea". International Journal of Pediatric Otorhinolaryngology. 76 (4): 471–4. 2012. doi:10.1016/j.ijporl.2011.12.026. PMID 22297210.

- Journal of Clinical Investigation. "Evidence of premature immune aging in patients thymectomized during early childhood". JCI. Retrieved 2012-06-11.

- "Human acute myeloid leukemia is organized as a hierarchy that originates from a primitive hematopoietic cell". Nature. 1997. Retrieved 9 November 2012.

5

Disorders of the Immune System

The immune system can fall prey to malfunctioning and disorders of the immune system fall into three broad categories – immunodeficiency, autoimmunity and hypersensitivities. The chapter explores each of these disorders and provides an integrated study of each. A section of the chapter specifically deals with transplant rejection, HIV and AIDS and chronic granulomatous disease.

Autoimmune Disease

An autoimmune disease is a pathological state arising from an abnormal immune response of the body to substances and tissues that are normally present in the body.

Autoimmunity, on the other hand, is the presence of self-reactive immune response (e.g., auto-antibodies, self-reactive T-cells), with or without damage or pathology resulting from it. This may be restricted to certain organs (e.g. in autoimmune thyroiditis) or involve a particular tissue in different places (e.g. Goodpasture's disease which may affect the basement membrane in both the lung and the kidney).

The treatment of autoimmune diseases is typically with immunosuppression—medication that decreases the immune response. New novel treatments include Cytokine blockade (or the blockade of cytokine signaling pathways), removal of effector T-cells and B-cells (e.g. anti-CD20 therapy can be effective at removing instigating B-cells). Intravenous Immunoglobulin has been helpful in treating some antibody mediated autoimmune diseases as well, possibly through negative feedback mechanisms.

A large number of autoimmune diseases are recognized. A major understanding of the underlying pathophysiology of autoimmune diseases has been the application of genome wide association scans that have identified a striking degree of genetic sharing among the autoimmune diseases.

Definition

For a disease to be regarded as an autoimmune disease it needs to answer to *Witebsky's postulates* (first formulated by Ernest Perez and colleagues in 1957 and modified in 1994):

- Direct evidence from transfer of disease-causing antibody or disease-causing T lymphocyte white blood cells
- Indirect evidence based on reproduction of the autoimmune disease in experimental animals
- Circumstantial evidence from clinical clues
- Genetic evidence suggesting "clustering" with other autoimmune diseases
- Autoimmune diseases are incurable

Effects

Autoimmune diseases have a wide variety of different effects. They do tend to have one of three characteristic pathological effects which characterize them as autoimmune diseases:

- Damage to or destruction of tissues

- Altered organ growth

- Altered organ function

It has been estimated that autoimmune diseases are among the leading causes of death among women in the United States in all age groups up to 65 years.

A substantial minority of the population suffers from these diseases, which are often chronic, debilitating, and life-threatening.

There are more than 80 illnesses caused by autoimmunity. Autoimmune diseases affect approximately 2-5% of the western world's population. Women are found to be more commonly affected than men. Environmental events can trigger some cases of autoimmune diseases such as exposure to radiation or certain drugs which can damage tissues of the body. Infection can also be a trigger of some autoimmune diseases for example Lupus which is thought to be a milder version of an idiopathic disorder where there is an increased production of antihistone antibodies.

Pathophysiology

The human immune system typically produces both T-cells and B-cells that are capable of being reactive with self-antigens, but these self-reactive cells are usually either killed prior to becoming active within the immune system, placed into a state of anergy (silently removed from their role within the immune system due to over-activation), or removed from their role within the immune system by regulatory cells. When any one of these mechanisms fail, it is possible to have a reservoir of self-reactive cells that become functional within the immune system. The mechanisms of preventing self-reactive T-cells from being created takes place through Negative selection process within the thymus as the T-cell is developing into a mature immune cell.

Some infections, such as Campylobacter jejuni, have antigens that are similar (but not identical) to our own self-molecules. In this case, a normal immune response to C. jejuni can result in the production of antibodies that also react to a lesser degree with receptors on skeletal muscle (i.e., Myasthenia gravis).

There are many theories as to how an autoimmune disease state arises. Some common ones are listed below.

Cryptic Determinants/Molecular Sequestration

Although it is possible for a potential auto antigen to be geographically sequestered in an immune privileged site within the body (e.g. the eye), mechanisms exist to express even these antigens in a tolerogenic fashion to the immune system. However, it is impossible to induce tolerance (immune unresponsiveness) to all aspects of an autoantigen. This is because under normal physiologic conditions some regions of a self-antigen are not expressed at a sufficient level to induce tolerance. These poorly displayed areas of an antigen are called "cryptic determinants." The immune system maintains a high-affinity repertoire to the cryptic self because the presentation of these determinants was insufficient to induce strong tolerance.

Molecular Mimicry

The concept of molecular mimicry describes a situation in which a foreign antigen can initiate an immune response in which a T or B cell component cross-recognizes self. The cross reactive immune response is responsible for the autoimmune disease state. Cross-reactive immune responses to self were first described for antibodies.

Altered Glycan Theory

According to this theory the effector function of the immune response is mediated by the glycans (polysaccharides) displayed by the cells and humoral components of the immune system. Individuals with autoimmunity have alterations in their glycosylation profile such that a proinflammatory immune response is favored. It is further hypothesized that individual autoimmune diseases will have unique glycan signatures.

Prevalence

The first estimate of US prevalence for autoimmune diseases as a group was published in 1997 by Jacobson, et al. They reported US prevalence to be around 9 million, applying prevalence estimates for 24 diseases to a US population of 279 million. Jacobson's work was updated by Hayter & Cook in 2012. This study used Witebsky's postulates, as revised by Rose & Bona, to extend the list to 81 diseases and estimated overall cumulative US prevalence for the 81 autoimmune diseases at 5.0%, with 3.0% for males and 7.1% for females. The estimated community prevalence, which takes into account the observation that many people have more than one autoimmune disease, was 4.5% overall, with 2.7% for males and 6.4% for females. Based on a US population of 320 million, the Hayter & Cook data indicate US prevalence at about 14.5 million.

In a report to the US Congress, the National Institutes of Health reported prevalence between 14.7 and 23.5 million. Although the NIH report predates the Hayter & Cook study by 7 years, the NIH low-range number closely matches their estimate. However, there is no substantiation provided for the higher estimate, and no list of diseases, or any reference to a list of diseases, is included in that report.

List by Category

This list of autoimmune diseases is categorized by organ and tissue type to help locate diseases that may be similar.

- Major Organs
  - Heart
  - Kidney

- • Liver
- • Lung
- • Skin
- Glands
  - • Endocrine
  - • Adrenal Gland
  - • Multi-glandular
  - • Pancreas
  - • Thyroid Gland
  - • Exocrine
  - • Reproductive Organs
  - • Salivary Glands
- Digestive System
- Tissue
  - • Blood
  - • Connective Tissue, Systemic, and multi-organ
  - • Muscle
  - • Nervous System
  - • Eyes
  - • Ears
  - • Vascular system
- Autoimmune Comorbidities
- Not Autoimmune

	Other Qualifiers
A	"Accepted" in prior version of this table
C	A disease, regarded as autoimmune, that is often found in individuals with another autoimmune condition. This designation is given to diseases that are classified by Rose Bona as having "circumstantial" evidence of autoimmune etiology. Diseases in this list with a "C" are, therefore, actual autoimmune diseases, rather than comorbid symptoms, which appear after this list.
E	Disease is an autoimmune response triggered by a specific environmental factor
F	Disease is only caused by autoimmunity in only a fraction of those who suffer from it

I	Described as an autoinflammatory disease
L	Evidence to indicate autoimmunity is extremely limited or circumstantial
M	Disease appears under Autoimmune Diseases in MeSH
N	Not listed in prior version of this table
R	Disease appeared in prior version but has been renamed. In renaming, precedence has been given to scientific names over those based on discoverers.
S	"Suspected" in the prior version of this table
T	Disease has a known trigger, such as viral infection, vaccination, or injury
X	An extremely rare disease, which would suggest limited opportunity to study it and conclusively determine whether it is caused by autoimmunity
Y	Listed in the prior version of this table with "Accepted/Suspected" left blank

Autoimmune Diseases

This is a dynamic list and may never be able to satisfy particular standards for completeness. You can help by expanding it with reliably sourced entries.

Organ/Tissue Type Disease Name	Level of Acceptance for Autoimmunity	Hypersensitivity (I,II,III,IV)	ICD-9 Codes	Notes/Autoantibodies/ Synonyms/Rare Variants
Major Organs<Top>				
Heart<Top>				
Myocarditis	Moderate, F, R, A		391.2422429.0	Synonyms: Autoimmune myocarditis, Autoimmune cardiomyopathy, Coxsackie myocarditis
Postmyocardial infarction syndrome	Limited, R, Y		411.0	Autoantibodies: Myocardial neo-antigens formed as a result of the MI. Synonym: Dressler's syndrome
Postpericardiotomy syndrome	Limited, N		429.4	
Subacute bacterial endocarditis (SBE)	Limited, Y	III	421.0	Autoantibodies: essential mixed cryoglobulinemia.
Kidney<Top>				
Anti-Glomerular Basement Membrane nephritis	Moderate, R, M, A	II	446.21	Autoantibodies: Anti-Basement Membrane Collagen Type IV Protein. Synonyms: Goodpastures Syndrome, Glomerulonephritis Type 1
Interstitial cystitis	Limited, S		595.1	Mast cells.
Lupus nephritis	Comorbidity, N		583.81	A comorbidity of Systemic Lupus Erythematosis.

Liver<Top>				
Autoimmune hepatitis	Moderate, A	cell-mediated	571.42	Autoantibodies: ANA and SMA, LKM-1, LKM-2 or LKM-3; antibodies against soluble liver antigen (anti-SLA, anti-LP) no autoantibodies detected (~20%). Synonym: Lupoid hepatitis
Primary biliary cirrhosis (PBC)	Moderate, A		571.6	Autoantibodies: Anti-p62, Anti-sp100, Anti-Mitochondrial(M2), Anti-Ro aka SSA. Note that Sjogren's is classified in some places (e.g. MeSH) as rheumatoid disease, but there is no published evidence to support that classification.
Primary sclerosing cholangitis	Limited, Y		576.1	Possible overlap with primary biliary cirrhosis. Autoantibodies: HLA-DR52a.
Lung<Top>				
Antisynthetase syndrome	Limited, Y		279.49	
Skin<Top>				
Alopecia Areata	Moderate, A		704.01	Autoantibodies: T-cells. Synonyms: Alopecia areata - Patchy, Totalis, Universalis
Autoimmune Angioedema	Limited, F, N		277.6 995.1	
Autoimmune progesterone dermatitis	Limited, X, A		279.49	
Autoimmune urticaria	Comorbidity, A		708	
Bullous pemphigoid	Moderate, Y		694.5	Autoantibodies: IgG autoantibodies targeting the type XVII collagen component of hemidesmosomes.
Cicatricial pemphigoid	Limited, R, X, Y		694.61	precipitates C3. Autoantibodies: anti-BP-1, anti-BP-2. Synonyms: Benign Mucosal Pemphigoid, Ocular cicatricial pemphigoid
Dermatitis herpetiformis	Moderate, C, Y		694.0	Autoantibodies: IgA Eosinophilia; anti-epidermal transglutaminase antibodies.
Discoid lupus erythematosus	Limited, Y	III	695.4	IL-2 and IFN-gamma.
Epidermolysis bullosa acquisita	Moderate, Y		694.8	COL7A1.
Erythema nodosum	Limited, F, Y		695.2	

Gestational pemphigoid	Limited, R, Y		646.8	Autoantibodies: IgG and C3 misdirected antibodies intended to protect the placenta.
Hidradenitis suppurativa	Limited, C, S		705.83	
Lichen planus	Limited, Y		697.0	
Lichen sclerosus	Limited, C, Y		701.0	
Linear IgA disease (LAD)	Moderate, Y		646.8	
Morphea	Limited, C, S		701.0	
Pemphigus vulgaris	Moderate, M, A	II	694.4	Autoantibodies: Anti-Desmoglein 3 eosinophilia.
Pityriasis lichenoides et varioliformis acuta	Limited, C		696.2	
Mucha-Habermann disease	Limited, C, Y		696.2	T-cells. Synonyms: Pityriasis lichenoides, varioliformis acuta
Psoriasis	Moderate, A	IV?	696	CD-8 T-cells, HLA-Cw6, IL-12b, IL-23b, TNFalpha, NF-κB.
Systemic scleroderma	Limited, R, S		710.1	COL1A2 and TGF-β1. Autoantibodies: Anti-nuclear antibodies, anti-centromere and anti-scl70/anti-topoisomerase antibodies. Synonyms: Diffuse cutaneous systemic sclerosis, Systemic sclerosis, Scleroderma
Vitiligo	Limited, C, S		709.01	NALP-1 RERE, PTPN22, LPP, IL2RA, GZMB, UBASH3A and C1QTNF6.
Glands<Top>				
Endocrine<Top>				
Adrenal Gland<Top>				
Addison's disease	Moderate, F, Y		255	Autoantibodies: 21 hydroxylase.
Multi-glandular<Top>				
Autoimmune polyendocrine syndrome (APS) type 1	Moderate, A	Unknown or multiple	258.1	Synonyms: Whitaker's Syndrome, Autoimmune polyendocrinopathy-candidiasis-ectodermal dystrophy (APECED), Addisons Disease, Polyglandular Autoimmune Syndrome 1 (PGAS-1).

Autoimmune polyendocrine syndrome (APS) type 2	Moderate, A		258.1	DQ2, DQ8 and DRB1*0404. Autoantibodies: anti-21 hydroxylase, anti-17 hydroxylase. Synonyms: Schmidt syndrome, Polyglandular Autoimmune Syndrome 2 (PGAS-2).
Autoimmune polyendocrine syndrome (APS) type 3	Moderate, A		258.1	Synonym: Polyglandular Autoimmune Syndrome 3 (PGAS-3).
Pancreas<Top>				
Autoimmune pancreatitis (AIP)	Moderate, A		577.1	Autoantibodies: ANA, anti-lactoferrin antibodies, anti-carbonic anhydrase antibodies, rheumatoid factor.
Diabetes mellitus type 1	Moderate, A	IV	250.01	HLA-DR3, HLA-DR4. Autoantibodies: Glutamic acid decarboxylase antibodies (GADA), islet cell antibodies (ICA), insulinoma-associated autoantibodies (IA-2), anti-insulin antibodies.
Thyroid Gland<Top>				
Autoimmune thyroiditis	Strong, A	IV	245.8	HLADR5, CTLA-4. Autoantibodies: Antibodies against thyroid peroxidase and/or thyroglobulin. Synonyms: Chronic lymphocytic thyroiditis, Hashimoto's thyroiditis.
Ord's thyroiditis	Moderate, Y		245.8	
Graves' disease	Moderate, M, A	II	242.0	Autoantibodies: Thyroid autoantibodies (TSHR-Ab) that activate the TSH-receptor (TSHR).
Exocrine<Top>				
Reproductive Organs<Top>				
Autoimmune Oophoritis	Moderate, N		614.2	
Endometriosis	Limited, S		617.0	
Autoimmune orchitis	Limited, N		604.0	
Salivary Glands<Top>				
Sjogren's syndrome	Moderate, A		710.2	Autoantibodies: Anti-Ro (often present also in Systemic Lupus Erythematosus).
Digestive System<Top>				
Autoimmune enteropathy	Moderate, X, Y			

Disease				
Coeliac disease	Moderate, A,E	IV??	579.0	HLA-DQ8 and DQ2.5. Autoantibodies: Anti-tissue transglutaminase antibodies, anti-endomysial IgA, anti-gliadin IgA.
Crohn's disease	Moderate, Y	IV	555	Innate immunity; Th17; Th1; ATG16L1; CARD15; XBP1.
Microscopic colitis	Limited, S		558.9	
Ulcerative colitis	Limited, A	IV	556	
Tissue<Top>				
Blood<Top>				
Antiphospholipid syndrome (APS, APLS)	Moderate, M, A		289.81	HLA-DR7, HLA-B8, HLA-DR2, HLA-DR3. Autoantibodies: Anti-cardiolipin; anti-pyruvate dehydrogenase; β2 glycoprotein I; phosphatidylserine; anti-apoH; Annexin A5. Synonym: Hughes syndrome.
Aplastic anemia	Limited, F, Y		284	
Autoimmune hemolytic anemia	Moderate, M, A	II	283.0	Complement activation.
Autoimmune lymphoproliferative syndrome	Moderate, A		279.41	TNFRSF6; defective Fas-CD95 apoptosis. Synonym: Canale-Smith syndrome.
Autoimmune neutropenia	Moderate, F, N		288.09	
Autoimmune thrombocytopenic purpura	Moderate, M, R, A		287.31	Autoantibodies: Anti gpIIb-IIIa or 1b-IX. Synonym: Idiopathic Thrombocytopenic Purpura (ITP)
Cold agglutinin disease	Moderate, M, A	II	283.0	Idiopathic or secondary to leukemia or infection. Autoantibodies: IgM. Synonym: Autoimmune hemolytic anemia.
Essential mixed cryoglobulinemia	Limited, C, Y		273.2	
Evans syndrome	Moderate, Y		287.32	Syndrome with a combination of hemolytic anemia and thrombocytopenic purpura
Paroxysmal nocturnal hemoglobinuria	Limited, F, S		283.2	
Pernicious anemia	Moderate, A	II	281.0	Autoantibodies: Anti-parietal cell antibody.
Pure red cell aplasia	Limited, Y		284.81	

Thrombocytopenia	Limited, F, Y	II	287.5	Multiple mechanisms. Autoantibodies: Glycoproteins IIb-IIIa or Ib-IX in ITP anti-ADAMTS13 in TTP. and HUS anti-cardiolipin (anti-cardiolipin antibodies) and β2 glycoprotein I in Antiphospholipid syndrome; anti-HPA-1a, anti-HPA-5b, and others in NAIT. Synonyms: Neonatal thrombocytopenia
Connective Tissue, Systemic, and multi-organ<Top>				
Adiposis dolorosa	Limited, L, S		272.8	Lipoid tissue. Synonym: Dercum's disease
Adult-onset Still's disease	Moderate, Y		714.2	Macrophage migration inhibitory factor. Autoantibodies: ANA.
Ankylosing Spondylitis	Limited, S		720.0	CD8; HLA-B27.
CREST syndrome	Limited, Y		710.1	Autoantibodies: Anti-centromere antibodies Anti-nuclear antibodies.
Drug-induced lupus	Moderate, Y		710.0	Autoantibodies: Anti-histone antibodies.
Enthesitis-related arthritis	Limited, C, Y			MMP3, TRLR2, TLR4, ERAP1. A subtype of Juvenile Rheumatoid Arthritis.
Eosinophilic fasciitis	Limited, F, A		728.89	Synonym: Shulman's syndrome
Felty syndrome	Strong, M, Y		714.1	
IgG4-related disease	Limited, C, N			Characteristic histological features (storiform fibrosis, lymphoplasmacytic infiltrate, obliterative phlebitis) are required for definitive diagnosis. Synonyms: IgG4-related systemic disease, IgG4-related sclerosing disease, IgG4-related systemic sclerosing disease, IgG4-related autoimmune disease, IgG4-associated multifocal systemic fibrosis, IgG4-associated disease, IgG4 syndrome, Hyper-IgG4 disease, Systemic IgG4-related plasmacytic syndrome.
Juvenile Arthritis	Strong, M, R, Y		714.30	Autoantibodies: inconsistent ANA, Rheumatoid factor. Synonyms: Juvenile rheumatoid arthritis, Juvenile idiopathic arthritis

Lyme disease (Chronic)	Limited, L, T, N		088.81	
Mixed connective tissue disease (MCTD)	Moderate, M, A		710.8	HLA-DR4. Autoantibodies: Anti-nuclear antibody, anti-U1-RNP.
Palindromic rheumatism	Limited, Y		719.3	Autoantibodies: Anti-cyclic citrullinated peptide antibodies (anti-CCP) and antikeratin antibodies (AKA). Synonym: Hench-Rosenberg syndrome.
Parry Romberg syndrome	Limited, Y		349.89351.8	Autoantibodies: ANA.
Parsonage-Turner syndrome	Limited, Y		353.5	
Psoriatic arthritis	Moderate, C, A	IV?	696.0	HLA-B27.
Reactive arthritis	Limited, C, F, Y		099.3	Synonym: Reiter's syndrome
Relapsing polychondritis	Strong, A		733.99	Synonyms: Atrophic polychondritis, systemic chondromalacia, chronic atrophic polychondritis, Meyenburg-Altherr-Uehlinger syndrome, generalized chondromalacia, systemic chondromalacia
Retroperitoneal fibrosis	Limited, Y		593.4	
Rheumatic fever	Moderate, T, A	II	390	Autoantibodies: Streptococcal M protein cross reacts with human myosin.
Rheumatoid arthritis	Strong, M, A	III	714	HLA-DR4, PTPN22, depleted B cells, TNF alpha, IL-17, (also maybe IL-1, 6, and 15). Autoantibodies: Rheumatoid factor (anti-IgGFc), Anti-MCV, ACPAs(Vimentin).
Sarcoidosis	Limited, S	IV	135	BTNL2; HLA-B7-DR15; HLA DR3-DQ2.
Schnitzler syndrome	Limited, L, X, Y		273.1	IgM.
Systemic Lupus Erythematosus (SLE)	Strong, M, A	III	695.4	Autoantibodies: Anti-nuclear antibodies, anti-Ro (often present also in Sjogren's syndrome). Eosinophilia. Synonym: Lupus
Undifferentiated connective tissue disease (UCTD)	Moderate, C, A		710.9	HLA-DR4. Autoantibodies: anti-nuclear antibody. Synonyms: Latent lupus, incomplete lupus

Muscle<Top>				
Dermatomyositis	Moderate, F, X, A		710.3	B- and T-cell perivascular inflammatory infiltrate on muscle biopsy. Autoantibodies: histidine-tRNA anti-signal recognition peptide Anti-Mi-2 Anti-Jo1. Synonym: Juvenile dermatomyositis
Fibromyalgia	Limited, C, F, N		729.1	
Inclusion body myositis	Limited, F, Y		359.71	Similar to polymyositis, but does not respond to steroid therapy-activated T8 cells.
Myositis	Limited, F, Y		729.1	
Myasthenia gravis	Strong, M, A	II	358	HA-B8 HLA-DR3 HLA-DR1. Autoantibodies: Nicotinic acetylcholine receptor MuSK protein.
Neuromyotonia	Limited, F, S	II?	333.90	Autoantibodies: Voltage-gated potassium channels. Synonym: Isaacs' syndrome
Paraneoplastic cerebellar degeneration	Limited, Y	IV? II?	334.9	Autoantibodies: anti-Yo (anti-cdr-2 in purkinje fibers) anti-Hu, anti-Tr, antiglutamate receptor.
Polymyositis	Limited, F, A		710.4	Autoantibodies: IFN-gamma, IL-1, TNF-alpha.
Nervous System<Top>				
Acute disseminated encephalomyelitis (ADEM)	Strong, M, T, A		323.61323.81	Synonyms: Perivenous encephalomyelitis, Acute hemorrhagic leukoencephalitis (AHL, AHLE), Acute necrotizing encephalopathy (ANE), Acute hemorrhagic encephalomyelitis (AHEM), Acute necrotizing hemorrhagic leukoencephalitis (ANHLE), Weston-Hurst syndrome, Hurst's disease.
Acute motor axonal neuropathy	Limited, N		356.8	
Anti-N-Methyl-D-Aspartate (Anti-NMDA) Receptor Encephalitis	Moderate, N			
Balo concentric sclerosis	Moderate, Y		341.1	Synonyms: Balo disease, Schilders disease.
Bickerstaff's encephalitis	Limited, Y		323.62	Similar to Guillain-Barré syndrome. Autoantibodies: Anti-GQ1b 2/3 patients.

Chronic inflammatory demyelinating polyneuropathy	Moderate, C, Y		357.81	Similar to Guillain–Barré syndrome. Autoantibodies: Anti-ganglioside antibodies. Synonyms: Relapsing polyneuropathy (CRP), chronic inflammatory demyelinating polyradiculoneuropathy, Chronic inflammatory demyelinating polyneuritis.
Guillain–Barré syndrome	Strong, M, A	IV	357.0	Autoantibodies: Anti-ganglioside, anti-GQ1b. Synonyms: Miller-Fisher syndrome, Landry's paralysis.
Hashimoto's encephalopathy	Moderate, C, X, A	IV		Autoantibodies: Alpha-enolase. Synonyms: Steroid-responsive encephalopathy associated with autoimmune thyroiditis (SREAT), Nonvasculitic autoimmune meningoencephalitis (NAIM), Encephalopathy Associated with Autoimmune Thyroid Disease (EAATD).
Idiopathic inflammatory demyelinating diseases	Limited, F, Y		356.8	A set of different variants of multiple sclerosis.
Lambert-Eaton myasthenic syndrome	Strong, M, Y		358.1	HLA-DR3-B8. Autoantibodies: Voltage-gated calcium channels; Q-type calcium channel, synaptogagmin, muscarinic acetylcholine receptor M1.
Multiple sclerosis, pattern II	Strong, M, A	IV	340	Autoantibody against potassium channel has been reported to present demyelination pattern II. Other cases present autoimmunity against MOG and Anoctamin-2. The three reported autoimmune variants belong to MS pattern II. Also involved HLA-DR2, PECAM-1, Anti-myelin basic protein. Autoantibodies: Anti-Kir4.1, Anti-MOG, Anti-ANO2 (heterogeneous). Synonyms: Primary progressive multiple sclerosis, Relapsing-remitting multiple sclerosis, disseminated sclerosis, encephalomyelitis disseminata.

Oshtoran Syndrome	X		F06.9	Heritable, abnormalities in the kynurenine and glutamate metabolism.
Pediatric Autoimmune Neuropsychiatric Disorder Associated with Streptococcus (PANDAS)	Limited, F, S	II?	279.49	Antibodies against streptococcal infection serve as auto-antibodies.
Progressive inflammatory neuropathy	Limited, X, S		356.4	Similar to Guillain-Barré syndrome. Autoantibodies: Anti-ganglioside antibodies:anti-GM1, anti-GD1a, anti-GQ1b.
Restless leg syndrome	Limited, C, S		333.94	May occur in Sjogren's syndrome, coeliac disease and rheumatoid arthritis, or in derangements of iron metabolism.
Stiff person syndrome	Limited, S		333.91	GLRA1 (glycine receptor). Autoantibodies: Glutamic acid decarboxylase (GAD).
Sydenham chorea	Limited, T, Y		392	
Transverse myelitis	Limited, M, A		323.82341.2	
Eyes<Top>				
Autoimmune retinopathy	Limited, X, N			
Autoimmune uveitis	Moderate, F, A		364	Autoantibodies: HLAB-27.
Cogan syndrome	Limited, F, Y		370.52	
Graves ophthalmopathy	Moderate, M, N		242.9	
Intermediate uveitis	Limited, L, Y		364.3	Synonyms: Pars planitis, Peripheral Uveitis.
Ligneous conjunctivitis	Limited, L, N		372.39	
Mooren's ulcer	Limited, L, N		370.07	
Neuromyelitis optica	Limited, M, Y	II?	341.0	Autoantibodies: NMO-IgG aquaporin 4. Synonym: Devic's disease.
Opsoclonus myoclonus syndrome	Limited, X, S	IV?	379.59	Lymphocyte recruitment to CSF.
Optic neuritis	Limited, C, Y		377.30	
Scleritis	Limited, C, Y		379.0	
Susac's syndrome	Limited, C, Y		348.39	Synonym: Retinocochleocerebral Vasculopathy.
Sympathetic ophthalmia	Limited, I, Y		360.11	Autoantibodies: ocular antigens following trauma.
Tolosa-Hunt syndrome	Limited, I, X, Y		378.55	

Ears<Top>				
Autoimmune inner ear disease (AIED)	Limited, A		388.8	
Ménière's disease	Limited, Y	III?	386.00	Autoantibodies: Major peripheral myelin protein Po.
Vascular system<Top>				
Behçet's disease	Limited, I, X, A		136.1	An immune-mediated systemic vasculitis; linkage to HLA-B51 (HLA-B27); very variable manifestations, with ulcers as common symptom. Synonyms: Morbus Adamandiades-Behçet. Rare Variant: Hughes-Stovin syndrome.
Eosinophilic granulomatosis with polyangiitis (EGPA)	Limited, I, X, Y		446.4	Autoantibodies: p-ANCA Eosinophilia. Synonym: Churg-Strauss syndrome.
Giant cell arteritis	Limited, I, R, A	IV	446.5	Synonyms: Cranial arteritis, Temporal Arteritis.
Granulomatosis with polyangiitis (GPA)	Strong, M, A		446.4	Autoantibodies: Anti-neutrophil cytoplasmic (cANCA). Synonym: Wegener's granulomatosis.
IgA vasculitis (IgAV)	Limited, L, Y		287.0	Autoantibodies: IgA and complement component 3 (C3). Synonyms: Anaphylactoid purpura, Henoch-Schonlein purpura, purpura rheumatica, Schönlein–Henoch purpura.
Kawasaki's disease	Moderate, S,E		446.1	ITPKC HLA-B51. Synonyms: Kawasaki syndrome, lymph node syndrome, mucocutaneous lymph node syndrome.
Leukocytoclastic vasculitis	Limited, L, Y		447.6	
Lupus vasculitis	Moderate, C, N		583.81	A comorbidity of Systemic Lupus Erythematosis.
Rheumatoid vasculitis	Moderate, C, N		447.6	A symptom of Lupus.
Microscopic polyangiitis (MPA)	Limited, Y		446.0	Binds to neutrophils causing them to degranulate and damages endothelium. Autoantibodies: p-ANCA myeloperoxidase. Synonyms: Microscopic polyarteritis, microscopic polyarteritis nodosa.

Polyarteritis nodosa (PAN)	Limited, L, Y		446.0	Synonyms: Panarteritis nodosa, periarteritis nodosa, Kussmaul disease, Kussmaul-Maier disease.
Polymyalgia rheumatica	Limited, L, Y		725	
Urticarial vasculitis	Limited, X, Y	II?	708.9	Clinically may resemble type I hypersensitivity. Autoantibodies: anti C1q antibodies.
Vasculitis	Strong, I, M, F, A	III	447.6	Autoantibodies: ANCA (sometimes).
Systemic<Top>				
Primary Immune Deficiency	Limited, N			

Autoimmune Comorbidities

This list includes conditions that are not diseases but signs common to autoimmune disease. Some, such as Chronic Fatigue Syndrome, are controversial. These conditions are included here because they are frequently listed as autoimmune diseases but should not be included in the list above until there is more consistent evidence.

Organ/Tissue Type Disease Name	Level of Acceptance for Autoimmunity	Hypersensitivity (I,II,III,IV)	ICD-9 Codes	Notes/Autoantibodies/Synonyms
Chronic fatigue syndrome	Comorbidity, N			Symptomatic of autoimmune diseases or autoimmune activity, but not a disease or a cause of disease.
Complex regional pain syndrome	Comorbidity, N			Symptomatic of autoimmune diseases or autoimmune activity, but not a disease or a cause of disease. Synonyms: Amplified Musculoskeletal Pain Syndrome, Reflex Neurovascular Dystrophy, Reflex sympathetic dystrophy
Eosinophilic esophagitis	Comorbidity, N		530.13	
Gastritis	Comorbidity, Y			Possibly symptomatic of autoimmune diseases, but not a disease or a cause of disease. Autoantibodies: serum antiparietal and anti-IF antibodies.
Interstitial lung disease	Comorbidity, N			Associated with several autoimmune connective tissue diseases.
POEMS syndrome	Comorbidity, Y			Possibly symptomatic of autoimmune diseases, but not a disease or a cause of disease. Autoantibodies: interleukin 1β, interleukin 6 and TNFα. vascular endothelial growth factor (VEGF), given the ..

Raynaud's phenomenon	Comorbidity, S			Symptomatic of autoimmune diseases or autoimmune activity, but not a disease or a cause of disease.
Primary immunodeficiency	Comorbidity, N		279.8	The condition is inherited, but it is associated with several autoimmune diseases.
Pyoderma gangrenosum	Comorbidity, Y			Possibly symptomatic of autoimmune diseases, but not a disease or a cause of disease.

Not Autoimmune

At this time, there is not sufficient evidence - direct, indirect, or circumstantial - to indicate that these diseases are caused by autoimmunity. These conditions are included here because:

1. The disease was listed in the prior version of this table

2. The disease is included in several widely used lists of autoimmune disease and is shown here to ensure that a person visiting this page does not conclude that the disease was not considered.

Organ/Tissue Type Disease Name	Level of Acceptance for Autoimmunity	Hypersensitivity (I,II,III,IV)	ICD-9 Codes	Notes/Autoantibodies/Synonyms
Agammaglobulinemia	Not Autoimmune, Y		279.00	An immune system disorder but not an autoimmune disease.. Autoantibodies: IGHM; IGLL1: CD79A; CD79B; BLNK; LRRC8A.
Amyloidosis	Not Autoimmune, N		277.30	No consistent evidence of association with autoimmunity.
Amyotrophic lateral sclerosis	Not Autoimmune, Y		335.20	No consistent evidence of association with autoimmunity. Autoantibodies: Amyotrophic lateral sclerosis (Also Lou Gehrig's disease; Motor Neuron Disease).
Anti-tubular basement membrane nephritis	Not Autoimmune, N			No consistent evidence of association with autoimmunity.
Atopic allergy	Not Autoimmune, Y	I	691.8	A hypersensitivity.
Atopic dermatitis	Not Autoimmune, Y	I	691.8	A hypersensitivity.
Autoimmune peripheral neuropathy	Not Autoimmune, F, A			A class of diseases, some of which may be autoimmune. See specific diseases that are listed as autoimmune..
Blau syndrome	Not Autoimmune, Y			Overlaps both sarcoidosis and granuloma annulare. No evidence of association with autoimmunity.

Cancer	Not Autoimmune, Y			No consistent evidence of association with autoimmunity.
Castleman's disease	Not Autoimmune, Y			An immune system disorder but not an autoimmune disease.. Autoantibodies: Over expression of IL-6.
Chagas disease	Not Autoimmune, S			No consistent evidence of association with autoimmunity.
Chronic obstructive pulmonary disease	Not Autoimmune, S			No consistent evidence of association with autoimmunity.
Chronic recurrent multifocal osteomyelitis	Not Autoimmune, Y			LPIN2, D18S60. Synonyms: Majeed syndrome
Complement component 2 deficiency	Not Autoimmune, Y			Possibly symptomatic of autoimmune diseases, but not a disease.
Congenital heart block	Not Autoimmune, N			May be related to autoimmune activity in the mother.
Contact dermatitis	Not Autoimmune, Y	IV		A hypersensitivity.
Cushing's syndrome	Not Autoimmune, Y			No consistent evidence of association with autoimmunity.
Cutaneous leukocytoclastic angiitis	Not Autoimmune, Y			No consistent evidence of association with autoimmunity. Autoantibodies: neutrophils.
Dego's disease	Not Autoimmune, Y			No consistent evidence of association with autoimmunity.
Eczema	Not Autoimmune, Y			No consistent evidence of association with autoimmunity. Autoantibodies: LEKTI, SPINK5, filaggrin., Brain-derived neurotrophic factor (BDNF) and Substance P..
Eosinophilic gastroenteritis	Not Autoimmune, Y			Possibly a hypersensitivity. Autoantibodies: IgE, IL-3, IL-5, GM-CSF, eotaxin.
Eosinophilic pneumonia	Not Autoimmune, F, Y			A class of diseases, some of which may be autoimmune. Specifically, Churg-Strauss syndrome, a subtype of Eosinophilic pneumonia, is autoimmune.
Erythroblastosis fetalis	Not Autoimmune, Y	II		Mother's immune system attacks fetus. An immune system disorder but not autoimmune. Autoantibodies: ABO, Rh, Kell antibodies.
Fibrodysplasia ossificans progressiva	Not Autoimmune, Y			Possibly an immune system disorder but not autoimmune. Autoantibodies: ACVR1 Lymphocytes express increased BMP4.

Gastrointestinal pemphigoid	Not Autoimmune, A			No consistent evidence of association with autoimmunity.
Hypogammaglobulinemia	Not Autoimmune, Y			An immune system disorder but not autoimmune. Autoantibodies: IGHM, IGLL1, CD79A, BLNK, LRRC8A, CD79B.
Idiopathic giant-cell myocarditis	Not Autoimmune, N			No consistent evidence of autoimmune cause though the disease has been found comorbid with other autoimmune diseases. Synonyms: Giant cell myocarditis
Idiopathic pulmonary fibrosis	Not Autoimmune, Y			Autoantibodies: SFTPA1, SFTPA2, TERT, and TERC.. Synonyms: Fibrosing alveolitis
IgA nephropathy	Not Autoimmune, Y	III?		Autoantibodies: IgA produced from marrow rather than MALT. Synonyms: IgA nephrits, Berger's disease, Synpharyngitic Glomerulonephritis. An immune system disorder but not an autoimmune disease.
Immunoregulatory lipoproteins	Not Autoimmune, N			Not a disease.
IPEX syndrome	Not Autoimmune, N			A genetic mutation in FOXP3 that leads to autoimmune diseases, but no consistent evidence that it is an autoimmune disorder itself.. Synonyms: X-linked polyendocrinopathy, immunodeficiency and diarrhea-syndrome (XLAAD)
Ligneous conjunctivitis	Not Autoimmune, N			No consistent evidence of association with autoimmunity.
Majeed syndrome	Not Autoimmune, Y			No consistent evidence of association with autoimmunity. Autoantibodies: LPIN2.
Narcolepsy	Not Autoimmune, Y	II?		No evidence of association with autoimmunity. Research not reproducible. Autoantibodies: hypocretin or orexin, HLA-DQB1*0602.
Rasmussen's encephalitis	Not Autoimmune, Y			No consistent evidence of association with autoimmunity. Autoantibodies: anti-NR2A antibodies.
Schizophrenia	Not Autoimmune, S			No consistent evidence of association with autoimmunity.
Serum sickness	Not Autoimmune, Y	III		A hypersensitivity.
Spondyloarthropathy	Not Autoimmune, Y			No consistent evidence of association with autoimmunity. Autoantibodies: HLA-B27.

Sweet's syndrome	Not Autoimmune, Y			No consistent evidence of association with autoimmunity. Autoantibodies: GCSF.
Takayasu's arteritis	Not Autoimmune, Y			No consistent evidence of association with autoimmunity.
Undifferentiated spondyloarthropathy	Not Autoimmune, Y			See Enthesitis-related arthritis.

Development of Therapies

In both autoimmune and inflammatory diseases, the condition arises through aberrant reactions of the human adaptive or innate immune systems. In autoimmunity, the patient's immune system is activated against the body's own proteins. In chronic inflammatory diseases, neutrophils and other leukocytes are constitutively recruited by cytokines and chemokines, leading to tissue damage.

Mitigation of inflammation by activation of anti-inflammatory genes and the suppression of inflammatory genes in immune cells is a promising therapeutic approach.

Hypersensitivity

Hypersensitivity (also called hypersensitivity reaction or intolerance) is a set of undesirable reactions produced by the normal immune system, including allergies and autoimmunity. These reactions may be damaging, uncomfortable, or occasionally fatal. Hypersensitivity reactions require a pre-sensitized (immune) state of the host. They are classified in four groups after the proposal of P. G. H. Gell and Robin Coombs in 1963.

Coombs and Gell Classification

Comparison of hypersensitivity types				
Type	Alternative names	Often mentioned disorders	Mediators	Description
I	Allergy (immediate)	• Atopy • Anaphylaxis • Asthma	• IgE	Fast response which occurs in minutes, rather than multiple hours or days. Free antigens cross link the IgE on mast cells and basophils which causes a release of vasoactive biomolecules. Testing can be done via skin test for specific IgE.

II	Cytotoxic, antibody-dependent	• Autoimmune hemo-lytic anemia • Rheumatic heart disease • Thrombocytopenia • Erythroblastosis fetalis • Goodpasture's syn-drome • Graves' disease *see type V explanation below • Myasthenia gravis *see type V explana-tion below	• IgM or IgG • (Complement) • MAC	Antibody (IgM or IgG) binds to antigen on a target cell, which is actually a host cell that is perceived by the immune system as foreign, leading to cellular destruction via the MAC. Testing includes both the direct and indirect Coombs test.
III	Immune complex disease	• Serum sickness • Arthus reaction • Rheumatoid arthritis • Post streptococcal glomerulonephritis • Membranous ne-phropathy • Lupus nephritis • Systemic lupus ery-thematosus • Extrinsic allergic alveolitis (hypersen-sitivity pneumonitis)	• IgG • (Complement) • Neutrophils	Antibody (IgG) binds to soluble antigen, forming a circulating immune complex. This is often deposited in the vessel walls of the joints and kidney, initiating a local inflammatory reaction.
IV	Delayed-type hypersensitivity, cell-mediated immune memory response, antibody-independent	• Contact dermatitis, including Urushi-ol-induced contact dermatitis (poison ivy rash). • Mantoux test • Chronic transplant rejection • Multiple sclerosis	• T-cells	Helper T cells (specifically Th1 helper t cells) are activated by an antigen presenting cell. When the antigen is presented again in the future, the memory Th1 cells will activate macrophages and cause an inflammatory response. This ultimately can lead to tissue damage.
V	Autoimmune disease, receptor mediated (see below)	• Graves' disease • Myasthenia gravis	• IgM or IgG • (Complement)	

Type V

This is an additional type that is sometimes (often in the UK) used as a distinction from Type 2.

Instead of binding to cell surfaces, the antibodies recognise and bind to the cell surface receptors, which either prevents the intended ligand binding with the receptor or mimics the effects of the ligand, thus impairing cell signaling.

Some clinical examples:

- Graves' disease

- Myasthenia gravis

The use of Type 5 is rare. These conditions are more frequently classified as Type 2, though sometimes they are specifically segregated into their own subcategory of Type 2.

Immunodeficiency

Immunodeficiency (or immune deficiency) is a state in which the immune system's ability to fight infectious disease is compromised or entirely absent. Most cases of immunodeficiency are acquired ("secondary") due to extrinsic factors that affect the patient's immune system. Examples of these extrinsic factors include infections, such as by Human Immunodeficiency virus (HIV), extremes of age and environmental factors, such as nutrition. In the clinical setting, the immuno-suppression quality of some drugs, such as steroids, can be utilised. Examples of such use is in transplant surgery as an anti-rejection measure and in patients suffering from an over-active immune system. Some people are born with defects in their immune system, or primary immuno-deficiency. A person who has an immunodeficiency of any kind is said to be immunocompromised. An immunocompromised person may be particularly vulnerable to opportunistic infections, in addition to normal infections that could affect everyone. Immunodeficiency may also decrease cancer immunosurveillance

Types

By Affected Component

- Humoral immune deficiency, with signs or symptoms depending on the cause, but generally include signs of hypogammaglobulinemia (decrease of one or more types of antibodies) with presentations including repeated mild respiratory infections, and/or agammaglobulinemia (lack of all or most antibody production) which results in frequent severe infections and is often fatal.

- T cell deficiency, often causes secondary disorders such as acquired immune deficiency syndrome (AIDS).

- *Granulocyte deficiency*, including decreased numbers of granulocytes (called as granulocytopenia or, if absent, agranulocytosis) such as of neutrophil granulocytes (termed neutropenia). Granulocyte deficiencies also include decreased function of individual granulocytes, such as in chronic granulomatous disease.

- Asplenia, where there is no function of the spleen

- Complement deficiency is where the function of the complement system is deficient

In reality, immunodeficiency often affects multiple components, with notable examples including severe combined immunodeficiency (which is primary) and acquired immune deficiency syndrome (which is secondary).

Comparison of immunodeficiencies by affected component			
	Affected components	Main causes	Main pathogens of resultant infections
Humoral immune deficiency	B cells, plasma cells or antibodies	• Primary humoral • Multiple myeloma • Chronic lymphoid leukemia • AIDS	• Streptococcus pneumoniae • Hemophilus influenzae • Pneumocystis jirovecii • Giardia intestinalis • Cryptosporidium parvum
T cell deficiency	T cells	• Marrow and other transplantation • AIDS • Cancer chemotherapy • Lymphoma • Glucocorticoid therapy	Intracellular pathogens, including *Herpes simplex virus*, *Mycobacterium*, *Listeria*, and intracellular fungal infections.
Neutropenia	Neutrophil granulocytes	• Chemotherapy • Bone marrow transplantation • Dysfunction, such as chronic granulomatous disease	• Enterobacteriaceae • Oral *Streptococci* • *Pseudomonas aeruginosa* • *Enterococcus* species • *Candida* species • *Aspergillus* species

Asplenia	Spleen	• Splenectomy
		• Trauma
		• Sickle-cell anemia
			• Polysaccharide encapsulated bacteria, particularly:
			  o *Streptococcus pneumoniae*
			  o *Haemophilus influenzae*
			  o *Neisseria meningitidis*
			• *Plasmodium* species
			• *Babesia* species

Complement deficiency	Complement system	• Congenital deficiencies	• *Neisseria* species • Streptococcus pneumoniae

Primary or Secondary

Distinction between primary versus secondary immunodeficiencies are based on, respectively, whether the cause originates in the immune system itself or is, in turn, due to insufficiency of a supporting component of it or an external decreasing factor of it.

Primary Immunodeficiency

A number of rare diseases feature a heightened susceptibility to infections from childhood onward. Primary Immunodeficiency is also known as congenital immunodeficiencies. Many of these disorders are hereditary and are autosomal recessive or X-linked. There are over 80 recognised primary immunodeficiency syndromes; they are generally grouped by the part of the immune system that is malfunctioning, such as lymphocytes or granulocytes.

The treatment of primary immunodeficiencies depends on the nature of the defect, and may involve antibody infusions, long-term antibiotics and (in some cases) stem cell transplantation.

Secondary Immunodeficiencies

Secondary immunodeficiencies, also known as acquired immunodeficiencies, can result from various immunosuppressive agents, for example, malnutrition, aging and particular medications (e.g., chemotherapy, disease-modifying antirheumatic drugs, immunosuppressive drugs after organ transplants, glucocorticoids). For medications, the term immunosuppression generally refers to both beneficial and potential adverse effects of decreasing the function of the immune system, while the term *immunodeficiency* generally refers solely to the adverse effect of increased risk for infection.

Many specific diseases directly or indirectly cause immunosuppression. This includes many types of cancer, particularly those of the bone marrow and blood cells (leukemia, lymphoma, multiple myeloma), and certain chronic infections. Immunodeficiency is also the hallmark of acquired immunodeficiency syndrome (AIDS), caused by the human immunodeficiency virus (HIV). HIV directly infects a small number of T helper cells, and also impairs other immune system responses indirectly.

Immunodeficiency and Autoimmunity

There are a large number of immunodeficiency syndromes that present clinical and laboratory characteristics of autoimmunity. The decreased ability of the immune system to clear infections in these patients may be responsible for causing autoimmunity through perpetual immune system activation.

One example is common variable immunodeficiency (CVID) where multiple autoimmune diseases are seen, e.g., inflammatory bowel disease, autoimmune thrombocytopenia and autoimmune thyroid disease. Familial hemophagocytic lymphohistiocytosis, an autosomal recessive primary immunodeficiency, is another example. Pancytopenia, rashes, lymphadenopathy and hepatosplenomegaly are commonly seen in these patients. Presence of multiple uncleared viral infections due to lack of perforin are thought to be responsible. In addition to chronic and/or recurrent infections many autoimmune diseases including arthritis, autoimmune hemolytic anemia, scleroderma and type 1 diabetes are also seen in X-linked agammaglobulinemia (XLA). Recurrent bacterial and fungal infections and chronic inflammation of the gut and lungs are seen in chronic granulomatous disease (CGD) as well. CGD is caused by a decreased production of nicotinamide adenine dinucleotide phosphate (NADPH) oxidase by neutrophils. Hypomorphic RAG mutations are seen in patients with midline granulomatous disease; an autoimmune disorder that is commonly seen in patients with granulomatosis with polyangiitis (Wegner's disease) and NK/T cell lymphomas. Wiskott-Aldrich syndrome (WAS) patients also present with eczema, autoimmune manifestations, recurrent bacterial infections and lymphoma. In autoimmune polyendocrinopathy-candidiasis-ectodermal dystrophy (APECED) also autoimmunity and infections coexist: organ-specific autoimmune manifestations (e.g., hypoparathyroidism and adrenocortical failure) and chronic mucocutaneous candidiasis. Finally, IgA deficiency is also sometimes associated with the development of autoimmune and atopic phenomena.

Management

Prevention of Pneumocystis pneumonia using trimethoprim/sulfamethoxazole is useful in those who are immunocompromised.

Transplant Rejection

Transplant rejection occurs when transplanted tissue is rejected by the recipient's immune system, which destroys the transplanted tissue. Transplant rejection can be lessened by determining the molecular similitude between donor and recipient and by use of immunosuppressant drugs after transplant.

Pretransplant Rejection Prevention

The first successful organ transplant, performed in 1954 by Joseph Murray, involved identical twins, and so no rejection was observed. Otherwise, the number of mismatched gene variants, namely alleles, encoding cell surface molecules called major histocompatibility complex (MHC), classes I and II, correlate with the rapidity and severity of transplant rejection. In humans MHC is also called human leukocyte antigen (HLA).

Though cytotoxic-crossmatch assay can predict rejection mediated by cellular immunity, genetic-expression tests specific to the organ type to be transplanted, for instance AlloMap Molecular Expression Testing, have a high negative predictive value. Transplanting only ABO-compatible grafts (matching blood groups between donor and recipient) helps prevent rejection mediated by humoral immunity.

ABO-Incompatible Transplants

Because very young children (generally under 12 months, but often as old as 24 months) do not have a well-developed immune system, it is possible for them to receive organs from otherwise incompatible donors. This is known as ABO-incompatible (ABOi) transplantation. Graft survival and patient mortality is approximately the same between ABOi and ABO-compatible (ABOc) recipients. While focus has been on infant heart transplants, the principles generally apply to other forms of solid organ transplantation.

The most important factors are that the recipient not have produced isohemagglutinins, and that they have low levels of T cell-independent antigens. UNOS regulations allow for ABOi transplantation in children under two years of age if isohemagglutinin titers are 1:4 or below, and if there is no matching ABOc recipient. Studies have shown that the period under which a recipient may undergo ABOi transplantation may be prolonged by exposure to nonself A and B antigens. Furthermore, should the recipient (for example, type B-positive with a type AB-positive graft) require eventual retransplantation, the recipient may receive a new organ of either blood type.

Limited success has been achieved in ABO-incompatible heart transplants in adults, though this requires that the adult recipients have low levels of anti-A or anti-B antibodies. Kidney transplantation is more successful, with similar long-term graft survival rates to ABOc transplants.

Immunologic Mechanisms of Rejection

Rejection is an adaptive immune response via cellular immunity (mediated by killer T cells inducing apoptosis of target cells) as well as humoral immunity (mediated by activated B cells secreting antibody molecules), though the action is joined by components of innate immune response (phagocytes and soluble immune proteins). Different types of transplanted tissues tend to favor different balances of rejection mechanisms.

Immunization

An animal's exposure to the antigens of a different member of the same or similar species is *allostimulation*, and the tissue is *allogenic*. Transplanted organs are often acquired from a cadaver (usually a host who had succumbed to trauma), whose tissues had already sustained ischemia or inflammation.

Dendritic cells (DCs), which are the primary antigen-presenting cells (APCs), of the donor tissue migrate to the recipient's peripheral lymphoid tissue (lymphoid follicles and lymph nodes), and present the donor's *self* peptides to the recipient's lymphocytes (immune cells residing in lymphoid tissues). Lymphocytes include two classes that enact adaptive immunity, also called specific immunity. Lymphocytes of specific immunity T cells—including the subclasses helper T cells and killer T cells—and B cells.

The recipient's helper T cells coordinate specific immunity directed at the donor's *self* peptides or at the donor's Major histocompatibility complex molecules, or at both.

Immune Memory

When memory helper T cells' CD4 receptors bind to the MHC class II molecules which are expressed on the surfaces of the target cells of the graft tissue, the memory helper T cells' T cell receptors (TCRs) can recognize their target antigen that is presented by the MHC class II molecules. The memory helper T cell subsequently produces clones that, as effector cells, secrete immune signalling molecules (cytokines) in approximately the cytokine balance that had prevailed at the memory helper T cell's priming to memorize the antigen. As the priming event in this instance occurred amid inflammation, the immune memory is pro-inflammatory.

Cellular Immunity

As a cell is indicated by the prefix *cyto*, a cytotoxic influence destroys the cell. Alloreactive killer T cells, also called cytotoxic T lymphocytes (CTLs), have CD8 receptors that dock to the transplanted tissue's MHC class I molecules,which display the donor's self peptides. (In the living donor, such presentation of *self* antigens helped maintain *self* tolerance.) Thereupon, the T cell receptors (TCRs) of the killer T cells recognize their matching epitope, and trigger the target cell's programmed cell death by apoptosis.

Humoral Immunity

Developed through an earlier *primary exposure* that primed specific immunity to the *nonself* antigen, a transplant recipient can have specific antibody crossreacting with the donor tissue upon the transplant event, a *secondary exposure*. This is typical after earlier mismatching among A/B/O blood types during blood transfusion. At this secondary exposure, these crossreactive antibody molecules interact with aspects of innate immunity—soluble immune proteins called complement and innate immune cells called phagocytes—which inflames and destroys the transplanted tissue.

Antibody

Secreted by an activated B cell, then called plasma cell, an antibody molecule is a soluble immunoglobulin (Ig) whose basic unit is shaped like the letter Y: the two arms are the Fab regions, while the single stalk is the Fc region. Each of the two tips of Fab region is the paratope, which binds a matching molecular sequence and its 3D shape (conformation), altogether called epitope, within the target antigen.

Opsonization

The IgG's Fc region also enables opsonization by a phagocyte, a process by which the Fc receptor on the phagocyte—such as neutrophils in blood and macrophages in tissues—binds the antibody molecule's FC stalk, and the phagocyte exhibits enhanced uptake of the antigen, attached to the antibody molecule's Fab region.

Complement Cascade

When the paratope of Ig class *gamma* (IgG) binds its matching epitope, IgG's Fc region conformationally shifts and can host a complement protein, initiating the complement cascade that terminates by punching a hole in a cell membrane. With many holes so punched, fluid rushes into the cell and ruptures it.

Cell debris can be recognized as damage associated molecular patterns (DAMPs) by pattern recognition receptors (PRRs), such as Toll-like receptors (TLRs), on membranes of phagocytes, which thereupon secrete proinflammatory cytokines, recruiting more phagocytes to traffic to the area by sensing the concentration gradient of the secreted cytokines (chemotaxis).

Tissue	Mechanism
Blood	Antibodies (isohaemagglutinins)
Kidney	Antibodies, cell-mediated immunity (CMI)
Heart	Antibodies, CMI

Skin	CMI
Bonemarrow	CMI
Cornea	Usually accepted unless vascularised: CMI

Medical Categories of Rejection

Hyperacute Rejection

Initiated by preexisting humoral immunity, *hyperacute rejection* manifests within minutes after transplant, and if tissue is left implanted brings systemic inflammatory response syndrome. Of high risk in kidney transplants is rapid clumping, namely agglutination, of red blood cells (RBCs or erythrocytes), as an antibody molecule binds multiple target cells at once.

While kidneys can routinely be obtained from human donors, most organs are in short supply leading to consideration of xenotransplants from other species. Pigs are especially likely sources for xenotransplants, chosen for the anatomical and physiological characteristics they share with humans. However, the sugar galactose-alpha-1,3-galactose (αGal) has been implicated as a major factor in hyperacute rejection in xenotransplantation. Unlike virtually all other mammals, humans and other primates do not make αGal, and in fact recognize it as an antigen. During transplantation, xenoreactive natural antibodies recognize αGal on the graft endothelium as an antigen, and the resulting complement-mediated immune response leads to a rejection of the transplant.

Acute Rejection

Developing with formation of cellular immunity, *acute rejection* occurs to some degree in all transplants, except between identical twins, unless immunosuppression is achieved (usually through drugs). Acute rejection begins as early as one week after transplant, the risk being highest in the first three months, though it can occur months to years later. Highly vascular tissues such as kidney or liver often host the earliest signs—particularly at endothelial cells lining blood vessels—though it eventually occurs in roughly 10 to 30% of liver transplants, and 10 to 20% of kidney transplants. A single episode of acute rejection can be recognized and promptly treated, usually preventing organ failure, but recurrent episodes lead to *chronic rejection*. It is believed that the process of acute rejection is mediated by the cell mediated pathway, specifically by mononuclear macrophages and T-lymphocytes.

Chronic Rejection

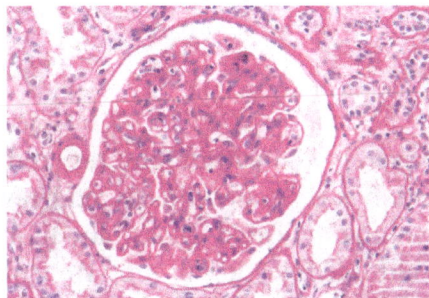

Micrograph showing a glomerulus with changes characteristic of a transplant glomerulopathy. Transplant glomerulopathy is considered a form of chronic antibody-mediated rejection. PAS stain.

The term *chronic rejection* initially described long-term loss of function in transplanted organs via fibrosis of the transplanted tissue's blood vessels. This is now *chronic allograft vasculopathy*, however, leaving *chronic rejection* referring to rejection due to more patent aspects of immunity.

Chronic rejection explains long-term morbidity in most lung-transplant recipients, the median survival roughly 4.7 years, about half the span versus other major organ transplants. In histopathology the condition is *bronchiolitis obliterans*, which clinically presents as progressive airflow obstruction, often involving dyspnea and coughing, and the patient eventually succumbs to pulmonary insufficiency or secondary acute infection.

Airflow obstruction not ascribable to other cause is labeled bronchiolitis obliterans syndrome (BOS), confirmed by a persistent drop—three or more weeks—in *forced expiratory volume* (FEV_1) by at least 20%. BOS is seen in over 50% of lung-transplant recipients by 5 years, and in over 80% by ten years. First noted is infiltration by lymphocytes, followed by epithelial cell injury, then inflammatory lesions and recruitment of fibroblasts and myofibroblasts, which proliferate and secrete proteins forming scar tissue. Generally thought unpredictable, BOS progression varies widely: lung function may suddenly fall but stabilize for years, or rapidly progress to death within a few months. Risk factors include prior acute rejection episodes, gastroesophageal reflux disease, acute infections, particular age groups, HLA mis-matching, lymphocytic bronchiolitis, and graft dysfunction (e.g., airway ischemia).

Rejection due to Non-Adherence

One principal reason for transplant rejection is non-adherence to prescribed immunosuppressant regimens. This is particularly the case with adolescent recipients, with non-adherence rates near 50% in some instances.

Rejection Detection

Diagnosis of acute rejection relies on clinical data—patient signs and symptoms—but also calls on laboratory data such as tissue biopsy. The laboratory pathologist generally seeks three main histological signs: (1) infiltrating T cells, perhaps accompanied by infiltrating eosinophils, plasma cells, and neutrophils, particularly in telltale ratios, (2) structural compromise of tissue anatomy, varying by tissue type transplanted, and (3) injury to blood vessels. Tissue biopsy is restricted, however, by sampling limitations and risks/complications of the invasive procedure. Cellular magnetic resonance imaging (MRI) of immune cells radiolabeled *in vivo* might offer noninvasive testing.

Rejection Treatment

Hyperacute rejection manifests severely and within minutes, and so treatment is immediate: removal of the tissue. Chronic rejection is generally considered irreversible and poorly amenable to treatment—only retransplant generally indicated if feasible—though inhaled ciclosporin is being investigated to delay or prevent chronic rejection of lung transplants. Acute rejection is treated with one or several of a few strategies.

Immunosuppressive Therapy

A short course of high-dose corticosteroids can be applied, and repeated. *Triple therapy* adds a calcineurin inhibitor and an anti-proliferative agent. Where calcineurin inhibitors or steroids are contraindicated, mTOR inhibitors are used.

Immunosuppressive Drugs:

- Corticosteroids

    o   Prednisolone

    o   Hydrocortisone

- Calcineurin inhibitors

    o   Ciclosporin

    o   Tacrolimus

- Anti-proliferatives

    o   Azathioprine

    o   Mycophenolic acid

- mTOR inhibitors

    o   Sirolimus

    o   Everolimus

Antibody-Based Treatments

Antibody specific to select immune components can be added to immunosuppressive therapy. The monoclonal anti-T cell antibody OKT3, once used to prevent rejection, and still occasionally used to treat severe acute rejection, has fallen into disfavor, as it commonly brings severe cytokine release syndrome and late post-transplant lymphoproliferative disorder. (OKT3 is available in the United Kingdom for named-patient use only.)

Antibody Drugs:

- Monoclonal anti-IL-2Rα receptor antibodies

    o   Basiliximab

    o   Daclizumab

- Polyclonal anti-T-cell antibodies

    o   Anti-thymocyte globulin (ATG)

    o   Anti-lymphocyte globulin (ALG)

- Monoclonal anti-CD20 antibodies

  o Rituximab

Blood Transfer

Cases refractory to immunosuppressive or antibody therapy are sometimes given blood transfusions—removing antibody molecules specific to the transplanted tissue.

Marrow Transplant

Bone marrow transplant can replace the transplant recipient's immune system with the donor's, and the recipient accepts the new organ without rejection. The marrow's hematopoietic stem cells—the reservoir of stem cells replenishing exhausted blood cells including white blood cells forming the immune system—must be of the individual who donated the organ or of an identical twin or a clone. There is a risk of graft-versus-host disease (GVHD), however, whereby mature lymphocytes entering with marrow recognize the new host tissues as foreign and destroy them.

Autoimmunity

Autoimmunity is the system of immune responses of an organism against its own healthy cells and tissues. Any disease that results from such an aberrant immune response is termed an autoimmune disease. Prominent examples include celiac disease, diabetes mellitus type 1, sarcoidosis, systemic lupus erythematosus (SLE), Sjögren's syndrome, eosinophilic granulomatosis with polyangiitis, Hashimoto's thyroiditis, Graves' disease, idiopathic thrombocytopenic purpura, Addison's disease, rheumatoid arthritis (RA), ankylosing spondylitis, polymyositis (PM), and dermatomyositis (DM). Autoimmune diseases are very often treated with steroids.

The misconception that an individual's immune system is totally incapable of recognizing *self* antigens is not new. Paul Ehrlich, at the beginning of the twentieth century, proposed the concept of *horror autotoxicus*, wherein a "normal" body does not mount an immune response against its own tissues. Thus, any autoimmune response was perceived to be abnormal and postulated to be connected with human disease. Now, it is accepted that autoimmune responses are an integral part of vertebrate immune systems (sometimes termed "natural autoimmunity"), normally prevented from causing disease by the phenomenon of immunological tolerance to self-antigens. Autoimmunity should not be confused with alloimmunity.

Low-Level Autoimmunity

While a high level of autoimmunity is unhealthy, a low level of autoimmunity may actually be beneficial. Taking the experience of a beneficial factor in autoimmunity further, one might hypothesize with intent to prove that autoimmunity is always a self-defense mechanism of the mammal system to survive. The system does not randomly lose the ability to distinguish between self and non-self, the attack on cells may be the consequence of cycling metabolic processes necessary to keep the blood chemistry in homeostasis.

Second, autoimmunity may have a role in allowing a rapid immune response in the early stages of an infection when the availability of foreign antigens limits the response (i.e., when there are few pathogens present). In their study, Stefanova et al. (2002) injected an anti-MHC Class II antibody into mice expressing a single type of MHC Class II molecule (H-2^b) to temporarily prevent CD4+ T cell-MHC interaction. Naive CD4+ T cells (those that have not encountered any antigens before) recovered from these mice 36 hours post-anti-MHC administration showed decreased responsiveness to the antigen pigeon cytochrome C peptide, as determined by Zap-70 phosphorylation, proliferation, and Interleukin-2 production. Thus Stefanova et al. (2002) demonstrated that self-MHC recognition (which, if too strong may contribute to autoimmune disease) maintains the responsiveness of CD4+ T cells when foreign antigens are absent. This idea of autoimmunity is conceptually similar to play-fighting. The play-fighting of young cubs (TCR and self-MHC) may result in a few scratches or scars (low-level-autoimmunity), but is beneficial in the long-term as it primes the young cub for proper fights in the future.

Immunological Tolerance

Pioneering work by Noel Rose and Ernst Witebsky in New York, and Roitt and Doniach at University College London provided clear evidence that, at least in terms of antibody-producing B lymphocytes, diseases such as rheumatoid arthritis and thyrotoxicosis are associated with loss of immunological tolerance, which is the ability of an individual to ignore "self", while reacting to "non-self". This breakage leads to the immune system's mounting an effective and specific immune response against self determinants. The exact genesis of immunological tolerance is still elusive, but several theories have been proposed since the mid-twentieth century to explain its origin.

Three hypotheses have gained widespread attention among immunologists:

- Clonal Deletion theory, proposed by Burnet, according to which self-reactive lymphoid cells are destroyed during the development of the immune system in an individual. For their work Frank M. Burnet and Peter B. Medawar were awarded the 1960 Nobel Prize in Physiology or Medicine "for discovery of acquired immunological tolerance".

- Clonal Anergy theory, proposed by Nossal, in which self-reactive T- or B-cells become inactivated in the normal individual and cannot amplify the immune response.

- Idiotype Network theory, proposed by Jerne, wherein a network of antibodies capable of neutralizing self-reactive antibodies exists naturally within the body.

In addition, two other theories are under intense investigation:

- Clonal Ignorance theory, according to which autoreactive T cells that are not represented in the thymus will mature and migrate to the periphery, where they will not encounter the appropriate antigen because it is inaccessible tissues. Consequently, auto-reactive B cells, that escape deletion, cannot find the antigen or the specific helper T-cell.

- Suppressor population or Regulatory T cell theory, wherein regulatory T-lymphocytes (commonly CD4$^+$FoxP3$^+$ cells, among others) function to prevent, downregulate, or limit autoaggressive immune responses in the immune system.

Tolerance can also be differentiated into "Central" and "Peripheral" tolerance, on whether or not the above-stated checking mechanisms operate in the central lymphoid organs (Thymus and Bone Marrow) or the peripheral lymphoid organs (lymph node, spleen, etc., where self-reactive B-cells may be destroyed). It must be emphasised that these theories are not mutually exclusive, and evidence has been mounting suggesting that all of these mechanisms may actively contribute to vertebrate immunological tolerance.

A puzzling feature of the documented loss of tolerance seen in spontaneous human autoimmunity is that it is almost entirely restricted to the autoantibody responses produced by B lymphocytes. Loss of tolerance by T cells has been extremely hard to demonstrate, and where there is evidence for an abnormal T cell response it is usually not to the antigen recognised by autoantibodies. Thus, in rheumatoid arthritis there are autoantibodies to IgG Fc but apparently no corresponding T cell response. In systemic lupus there are autoantibodies to DNA, which cannot evoke a T cell response, and limited evidence for T cell responses implicates nucleoprotein antigens. In Celiac disease there are autoantibodies to tissue transglutaminase but the T cell response is to the foreign protein gliadin. This disparity has led to the idea that human autoimmune disease is in most cases (with probable exceptions including type I diabetes) based on a loss of B cell tolerance which makes use of normal T cell responses to foreign antigens in a variety of aberrant ways.

Immunodeficiency and Autoimmunity

There are a large number of immunodeficiency syndromes that present clinical and laboratory characteristics of autoimmunity. The decreased ability of the immune system to clear infections in these patients may be responsible for causing autoimmunity through perpetual immune system activation.

One example is common variable immunodeficiency (CVID) where multiple autoimmune diseases are seen, e.g. inflammatory bowel disease, autoimmune thrombocytopenia and autoimmune thyroid disease. Familial hemophagocytic lymphohistiocytosis, an autosomal recessive primary immunodeficiency, is another example. Pancytopenia, rashes, swollen lymph nodes and enlargement of the liver and spleen are commonly seen in such individuals. Presence of multiple uncleared viral infections due to lack of perforin are thought to be responsible. In addition to chronic and/or recurrent infections many autoimmune diseases including arthritis, autoimmune hemolytic anemia, scleroderma and type 1 diabetes mellitus are also seen in X-linked agammaglobulinemia (XLA). Recurrent bacterial and fungal infections and chronic inflammation of the gut and lungs are seen in chronic granulomatous disease (CGD) as well. CGD is a caused by decreased production of nicotinamide adenine dinucleotide phosphate (NADPH) oxidase by neutrophils. Hypomorphic RAG mutations are seen in patients with midline granulomatous disease; an autoimmune disorder that is commonly seen in patients with granulomatosis with polyangiitis (formerly known as Wegener's granulomatosis) and NK/T cell lymphomas. Wiskott-Aldrich syndrome (WAS) patients also present with eczema, autoimmune manifestations, recurrent bacterial infections and lymphoma. In autoimmune polyendocrinopathy-candidiasis-ectodermal dystrophy (APECED) also autoimmunity and infections coexist: organ-specific autoimmune manifestations (e.g. hypoparathyroidism and adrenocortical failure) and chronic mucocutaneous candidiasis. Finally, IgA deficiency is also sometimes associated with the development of autoimmune and atopic phenomena.

Genetic Factors

Certain individuals are genetically susceptible to developing autoimmune diseases. This susceptibility is associated with multiple genes plus other risk factors. Genetically predisposed individuals do not always develop autoimmune diseases.

Three main sets of genes are suspected in many autoimmune diseases. These genes are related to:

- Immunoglobulins
- T-cell receptors
- The major histocompatibility complexes (MHC).

The first two, which are involved in the recognition of antigens, are inherently variable and susceptible to recombination. These variations enable the immune system to respond to a very wide variety of invaders, but may also give rise to lymphocytes capable of self-reactivity.

Scientists such as Hugh McDevitt, G. Nepom, J. Bell and J. Todd have also provided strong evidence to suggest that certain MHC class II allotypes are strongly correlated with

- HLA DR2 is strongly positively correlated with Systemic Lupus Erythematosus, narcolepsy and multiple sclerosis, and negatively correlated with DM Type 1.
- HLA DR3 is correlated strongly with Sjögren's syndrome, myasthenia gravis, SLE, and DM Type 1.
- HLA DR4 is correlated with the genesis of rheumatoid arthritis, Type 1 diabetes mellitus, and pemphigus vulgaris.

Fewer correlations exist with MHC class I molecules. The most notable and consistent is the association between HLA B27 and spondyloarthropathies like ankylosing spondylitis and reactive arthritis. Correlations may exist between polymorphisms within class II MHC promoters and autoimmune disease.

The contributions of genes outside the MHC complex remain the subject of research, in animal models of disease (Linda Wicker's extensive genetic studies of diabetes in the NOD mouse), and in patients (Brian Kotzin's linkage analysis of susceptibility to SLE).

Recently, PTPN22 has been associated with multiple autoimmune diseases including Type I diabetes, rheumatoid arthritis, systemic lupus erythematosus, Hashimoto's thyroiditis, Graves' disease, Addison's disease, Myasthenia Gravis, vitiligo, systemic sclerosis juvenile idiopathic arthritis, and psoriatic arthritis.

Sex

Ratio of female/male incidence of autoimmune diseases	
Hashimoto's thyroiditis	10/1
Graves' disease	7/1

Multiple sclerosis (MS)	2/1
Myasthenia gravis	2/1
Systemic lupus erythematosus (SLE)	9/1
Rheumatoid arthritis	5/2
Primary sclerosing cholangitis	1/2

A person's sex also seems to have some role in the development of autoimmunity; that is, most autoimmune diseases are *sex-related*. Nearly 75% of the more than 23.5 million Americans who suffer from autoimmune disease are women, although it is less-frequently acknowledged that millions of men also suffer from these diseases. According to the American Autoimmune Related Diseases Association (AARDA), autoimmune diseases that develop in men tend to be more severe. A few autoimmune diseases that men are just as or more likely to develop as women include: ankylosing spondylitis, type 1 diabetes mellitus, granulomatosis with polyangiitis, Crohn's disease, Primary sclerosing cholangitis and psoriasis.

The reasons for the sex role in autoimmunity are unclear. Women appear to generally mount larger inflammatory responses than men when their immune systems are triggered, increasing the risk of autoimmunity. Involvement of sex steroids is indicated by that many autoimmune diseases tend to fluctuate in accordance with hormonal changes, for example, during pregnancy, in the menstrual cycle, or when using oral contraception. A history of pregnancy also appears to leave a persistent increased risk for autoimmune disease. It has been suggested that the slight exchange of cells between mothers and their children during pregnancy may induce autoimmunity. This would tip the gender balance in the direction of the female.

Another theory suggests the female high tendency to get autoimmunity is due to an imbalanced X chromosome inactivation. The X-inactivation skew theory, proposed by Princeton University's Jeff Stewart, has recently been confirmed experimentally in scleroderma and autoimmune thyroiditis. Other complex X-linked genetic susceptibility mechanisms are proposed and under investigation.

Environmental Factors

An interesting inverse relationship exists between infectious diseases and autoimmune diseases. In areas where multiple infectious diseases are endemic, autoimmune diseases are quite rarely seen. The reverse, to some extent, seems to hold true. The hygiene hypothesis attributes these correlations to the immune manipulating strategies of pathogens. Whilst such an observation has been variously termed as spurious and ineffective, according to some studies, parasite infection is associated with reduced activity of autoimmune disease.

The putative mechanism is that the parasite attenuates the host immune response in order to protect itself. This may provide a serendipitous benefit to a host that also suffers from autoimmune disease. The details of parasite immune modulation are not yet known, but may include secretion of anti-inflammatory agents or interference with the host immune signaling.

A paradoxical observation has been the strong association of certain microbial organisms with autoimmune diseases. For example, *Klebsiella pneumoniae* and coxsackievirus B have been strongly correlated with ankylosing spondylitis and diabetes mellitus type 1, respectively. This has been explained by the tendency of the infecting organism to produce super-antigens that are capable of polyclonal activation of B-lymphocytes, and production of large amounts of antibodies of varying specificities, some of which may be self-reactive.

Certain chemical agents and drugs can also be associated with the genesis of autoimmune conditions, or conditions that simulate autoimmune diseases. The most striking of these is the drug-induced lupus erythematosus. Usually, withdrawal of the offending drug cures the symptoms in a patient.

Cigarette smoking is now established as a major risk factor for both incidence and severity of rheumatoid arthritis. This may relate to abnormal citrullination of proteins, since the effects of smoking correlate with the presence of antibodies to citrullinated peptides.

Pathogenesis of Autoimmunity

Several mechanisms are thought to be operative in the pathogenesis of autoimmune diseases, against a backdrop of genetic predisposition and environmental modulation. It is beyond the scope of this article to discuss each of these mechanisms exhaustively, but a summary of some of the important mechanisms have been described:

- T-Cell Bypass – A normal immune system requires the activation of B-cells by T-cells before the former can undergo differentiation into plasma B-cells and subsequently produce antibodies in large quantities. This requirement of a T-cell can be bypassed in rare instances, such as infection by organisms producing super-antigens, which are capable of initiating polyclonal activation of B-cells, or even of T-cells, by directly binding to the β-subunit of T-cell receptors in a non-specific fashion.

- T-Cell-B-Cell discordance – A normal immune response is assumed to involve B and T cell responses to the same antigen, even if we know that B cells and T cells recognise very different things: conformations on the surface of a molecule for B cells and pre-processed peptide fragments of proteins for T cells. However, there is nothing as far as we know that requires this. All that is required is that a B cell recognising antigen X endocytoses and processes a protein Y (normally =X) and presents it to a T cell. Roosnek and Lanzavecchia showed that B cells recognising IgGFc could get help from any T cell responding to an antigen co-endocytosed with IgG by the B cell as part of an immune complex. In coeliac disease it seems likely that B cells recognising tissue transglutamine are helped by T cells recognising gliadin.

- Aberrant B cell receptor-mediated feedback – A feature of human autoimmune disease is that it is largely restricted to a small group of antigens, several of which have known signaling roles in the immune response (DNA, C1q, IgGFc, Ro, Con. A receptor, Peanut agglutinin receptor(PNAR)). This fact gave rise to the idea that spontaneous autoimmunity may result when the binding of antibody to certain antigens leads to aberrant signals being fed back to parent B cells through membrane bound ligands. These ligands include B cell receptor (for antigen), IgG Fc receptors, CD21, which binds complement C3d, Toll-like receptors 9 and 7 (which can bind DNA and nucleoproteins) and PNAR. More indirect aberrant activation of B cells can also be envisaged with autoantibodies to acetyl choline receptor (on thymic myoid cells) and hormone and hormone binding proteins. Together with the concept of T-cell-B-cell discordance this idea forms the basis of the hypothesis of self-perpetuating autoreactive B cells. Autoreactive B cells in spontaneous autoimmunity are seen as surviving because of subversion both of the T cell help pathway and of the feedback signal through B cell receptor, thereby overcoming the negative signals responsible for B cell self-tolerance without necessarily requiring loss of T cell self-tolerance.

- Molecular Mimicry – An exogenous antigen may share structural similarities with certain host antigens; thus, any antibody produced against this antigen (which mimics the self-antigens) can also, in theory, bind to the host antigens, and amplify the immune response. The idea of molecular mimicry arose in the context of Rheumatic Fever, which follows infection with Group A beta-haemolytic streptococci. Although rheumatic fever has been attributed to molecular mimicry for half a century no antigen has been formally identified (if anything too many have been proposed). Moreover, the complex tissue distribution of the disease (heart, joint, skin, basal ganglia) argues against a cardiac specific antigen. It remains entirely possible that the disease is due to e.g. an unusual interaction between immune complexes, complement components and endothelium.

- Idiotype Cross-Reaction – Idiotypes are antigenic epitopes found in the antigen-binding portion (Fab) of the immunoglobulin molecule. Plotz and Oldstone presented evidence that autoimmunity can arise as a result of a cross-reaction between the idiotype on an antiviral antibody and a host cell receptor for the virus in question. In this case, the host-cell receptor is envisioned as an internal image of the virus, and the anti-idiotype antibodies can react with the host cells.

- Cytokine Dysregulation – Cytokines have been recently divided into two groups according to the population of cells whose functions they promote: Helper T-cells type 1 or type 2. The second category of cytokines, which include IL-4, IL-10 and TGF-β (to name a few), seem to have a role in prevention of exaggeration of pro-inflammatory immune responses.

- Dendritic cell apoptosis – immune system cells called dendritic cells present antigens to active lymphocytes. Dendritic cells that are defective in apoptosis can lead to inappropriate systemic lymphocyte activation and consequent decline in self-tolerance.

- Epitope spreading or epitope drift – when the immune reaction changes from targeting the primary epitope to also targeting other epitopes. In contrast to molecular mimicry, the other epitopes need not be structurally similar to the primary one.

- Epitope modification or Cryptic epitope exposure – this mechanism of autoimmune disease is unique in that it does not result from a defect in the hematopoietic system. Instead, disease results from the exposure of cryptic N-glycan (polysaccharide) linkages common to lower eukaryotes and prokaryotes on the glycoproteins of mammalian non-hematopoietic cells and organs This exposure of phylogenically primitive glycans activates one or more mammalian innate immune cell receptors to induce a chronic sterile inflammatory state. In the presence of chronic and inflammatory cell damage, the adaptive immune system is recruited and self–tolerance is lost with increased autoantibody production. In this form of the disease, the absence of lymphocytes can accelerate organ damage, and intravenous IgG administration can be therapeutic. Although this route to autoimmune disease may underlie various degenerative disease states, no diagnostics for this disease mechanism exist at present, and thus its role in human autoimmunity is currently unknown.

The roles of specialized immunoregulatory cell types, such as regulatory T cells, NKT cells, γδ T-cells in the pathogenesis of autoimmune disease are under investigation.

Classification

Autoimmune diseases can be broadly divided into systemic and organ-specific or localised autoimmune disorders, depending on the principal clinico-pathologic features of each disease.

- Systemic autoimmune diseases include SLE, Sjögren's syndrome, sarcoidosis, scleroderma, rheumatoid arthritis, cryoglobulinemic vasculitis, and dermatomyositis. These conditions tend to be associated with autoantibodies to antigens which are not tissue specific. Thus although polymyositis is more or less tissue specific in presentation, it may be included in this group because the autoantigens are often ubiquitous t-RNA synthetases.

- Local syndromes which affect a specific organ or tissue:

    o Endocrinologic: Diabetes mellitus type 1, Hashimoto's thyroiditis, Addison's disease

    o Gastrointestinal: Coeliac disease, Crohn's Disease, Pernicious anaemia

    o Dermatologic: Pemphigus vulgaris, Vitiligo

    o Haematologic: Autoimmune haemolytic anaemia, Idiopathic thrombocytopenic purpura

    o Neurological: Myasthenia gravis

Using the traditional "organ specific" and "non-organ specific" classification scheme, many diseases have been lumped together under the autoimmune disease umbrella. However, many chronic inflammatory human disorders lack the telltale associations of B and T cell driven immunopathology. In the last decade it has been firmly established that tissue "inflammation against self" does not necessarily rely on abnormal T and B cell responses.

This has led to the recent proposal that the spectrum of autoimmunity should be viewed along an "immunological disease continuum," with classical autoimmune diseases at one extreme and diseases driven by the innate immune system at the other extreme. Within this scheme, the full spectrum of autoimmunity can be included. Many common human autoimmune diseases can be seen to have a substantial innate immune mediated immunopathology using this new scheme. This new classification scheme has implications for understanding disease mechanisms and for therapy development.

Diagnosis

Diagnosis of autoimmune disorders largely rests on accurate history and physical examination of the patient, and high index of suspicion against a backdrop of certain abnormalities in routine laboratory tests (example, elevated C-reactive protein). In several systemic disorders, serological assays which can detect specific autoantibodies can be employed. Localised disorders are best diagnosed by immunofluorescence of biopsy specimens. Autoantibodies are used to diagnose many autoimmune diseases. The levels of autoantibodies are measured to determine the progress of the disease.

Treatments

Treatments for autoimmune disease have traditionally been immunosuppressive, anti-inflammatory, or palliative. Managing inflammation is critical in autoimmune diseases. Non-immunological therapies, such as hormone replacement in Hashimoto's thyroiditis or Type 1 diabetes mellitus treat outcomes of the autoaggressive response, thus these are palliative treatments. Dietary manipulation limits the severity of celiac disease. Steroidal or NSAID treatment limits inflammatory symptoms of many diseases. IVIG is used for CIDP and GBS. Specific immunomodulatory therapies, such as the TNFα antagonists (e.g. etanercept), the B cell depleting agent rituximab, the anti-IL-6 receptor tocilizumab and the costimulation blocker abatacept have been shown to be useful in treating RA. Some of these immunotherapies may be associated with increased risk of adverse effects, such as susceptibility to infection.

Helminthic therapy is an experimental approach that involves inoculation of the patient with specific parasitic intestinal nematodes (helminths). There are currently two closely related treatments available, inoculation with either Necator americanus, commonly known as hookworms, or Trichuris Suis Ova, commonly known as Pig Whipworm Eggs.

T cell vaccination is also being explored as a possible future therapy for autoimmune disorders.

Nutrition and Autoimmunity

Vitamin D/Sunlight

- Because most human cells and tissues have receptors for vitamin D, including T and B cells, adequate levels of vitamin D can aid in the regulation of the immune system.

Omega-3 Fatty Acids

- Studies have shown that adequate consumption of omega-3 fatty acids counteracts the effects of arachidonic acids, which contribute to symptoms of autoimmune diseases. Human and animal trials suggest that omega-3 is an effective treatment modality for many cases of Rheumatoid Arthritis, Inflammatory Bowel Disease, Asthma, and Psoriasis.

- While major depression is not necessarily an autoimmune disease, some of is physiological symptoms are inflammatory and autoimmune in nature. Omega-3 may inhibit production of interferon gamma and other cytokines which cause the physiological symptoms of depression. This may be due to the fact that an imbalance in omega-3 and omega-6 fatty acids, which have opposing effects, is instrumental in the etiology of major depression.

Probiotics/Microflora

- Various types of bacteria and microflora present in fermented dairy products, especially *Lactobacillus casei*, have been shown to both stimulate immune response to tumors in mice and to regulate immune function, delaying or preventing the onset of nonobese diabetes. This is particularly true of the Shirota strain of *L. casei* (LcS). The LcS strain is mainly found in yogurt and similar products in Europe and Japan, and rarely elsewhere.

Antioxidants

- It has been theorized that free radicals contribute to the onset of type-1 diabetes in infants and young children, and therefore that the risk could be reduced by high intake of antioxidant substances during pregnancy. However, a study conducted in a hospital in Finland from 1997-2002 concluded that there was no statistically significant correlation between antioxidant intake and diabetes risk. It should be noted that this study involved monitoring of food intake through questionnaires, and estimated antioxidant intake on this basis, rather than by exact measurements or use of supplements.

HIV/AIDS

Human immunodeficiency virus infection and acquired immune deficiency syndrome (HIV/AIDS) is a spectrum of conditions caused by infection with the human immunodeficiency virus (HIV). Following initial infection, a person may not notice any symptoms or may experience a brief period of influenza-like illness. Typically, this is followed by a prolonged period with no symptoms. As the infection progresses, it interferes more with the immune system, increasing the risk of common infections like tuberculosis, as well as other opportunistic infections, and tumors that rarely affect people who have working immune systems. These late symptoms of infection are referred to as AIDS. This stage is often also associated with weight loss.

HIV is spread primarily by unprotected sex (including anal and oral sex), contaminated blood transfusions, hypodermic needles, and from mother to child during pregnancy, delivery, or breastfeeding. Some bodily fluids, such as saliva and tears, do not transmit HIV. Methods of prevention include safe sex, needle exchange programmes, treating those who are infected, and male circumcision. Disease in a baby can often be prevented by giving both the mother and child antiretroviral medication. There is no cure or vaccine; however, antiretroviral treatment can slow the course of the disease and may lead to a near-normal life expectancy. Treatment is recommended as soon as the diagnosis is made. Without treatment, the average survival time after infection is 11 years.

In 2014 about 36.9 million people were living with HIV and it resulted in 1.2 million deaths. Most of those infected live in sub-Saharan Africa. Between its discovery and 2014 AIDS has caused an estimated 39 million deaths worldwide. HIV/AIDS is considered a pandemic—a disease outbreak which is present over a large area and is actively spreading. HIV is believed to have originated in west-central Africa during the late 19th or early 20th century. AIDS was first recognized by the United States Centers for Disease Control and Prevention (CDC) in 1981 and its cause—HIV infection—was identified in the early part of the decade.

HIV/AIDS has had a great impact on society, both as an illness and as a source of discrimination. The disease also has large economic impacts. There are many misconceptions about HIV/AIDS such as the belief that it can be transmitted by casual non-sexual contact. The disease has become subject to many controversies involving religion including the Catholic church's decision not to support condom use as prevention. It has attracted international medical and political attention as well as large-scale funding since it was identified in the 1980s.

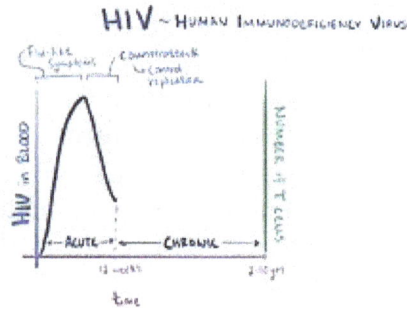

Video explanation

Signs and Symptoms

There are three main stages of HIV infection: acute infection, clinical latency and AIDS.

Acute Infection

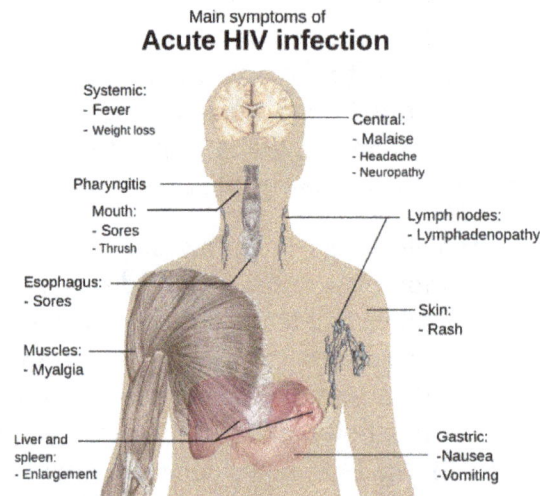

Main symptoms of acute HIV infection

The initial period following the contraction of HIV is called acute HIV, primary HIV or acute retro-viral syndrome. Many individuals develop an influenza-like illness or a mononucleosis-like illness 2–4 weeks post exposure while others have no significant symptoms. Symptoms occur in 40–90% of cases and most commonly include fever, large tender lymph nodes, throat inflammation, a rash, headache, and/or sores of the mouth and genitals. The rash, which occurs in 20–50% of cases, presents itself on the trunk and is maculopapular, classically. Some people also develop opportunistic infections at this stage. Gastrointestinal symptoms such as nausea, vomiting or diarrhea may occur, as may neurological symptoms of peripheral neuropathy or Guillain-Barre syndrome. The duration of the symptoms varies, but is usually one or two weeks.

Due to their nonspecific character, these symptoms are not often recognized as signs of HIV infection. Even cases that do get seen by a family doctor or a hospital are often misdiagnosed as one of the many common infectious diseases with overlapping symptoms. Thus, it is recommended that HIV be considered in people presenting an unexplained fever who may have risk factors for the infection.

Clinical Latency

The initial symptoms are followed by a stage called clinical latency, asymptomatic HIV, or chronic HIV. Without treatment, this second stage of the natural history of HIV infection can last from about three years to over 20 years (on average, about eight years). While typically there are few or no symptoms at first, near the end of this stage many people experience fever, weight loss, gastrointestinal problems and muscle pains. Between 50 and 70% of people also develop persistent generalized lymphadenopathy, characterized by unexplained, non-painful enlargement of more than one group of lymph nodes (other than in the groin) for over three to six months.

Although most HIV-1 infected individuals have a detectable viral load and in the absence of treatment will eventually progress to AIDS, a small proportion (about 5%) retain high levels of CD4$^+$ T cells (T helper cells) without antiretroviral therapy for more than 5 years. These individuals are classified as HIV controllers or long-term nonprogressors (LTNP). Another group consists of those who maintain a low or undetectable viral load without anti-retroviral treatment, known as "elite controllers" or "elite suppressors". They represent approximately 1 in 300 infected persons.

Acquired Immunodeficiency Syndrome

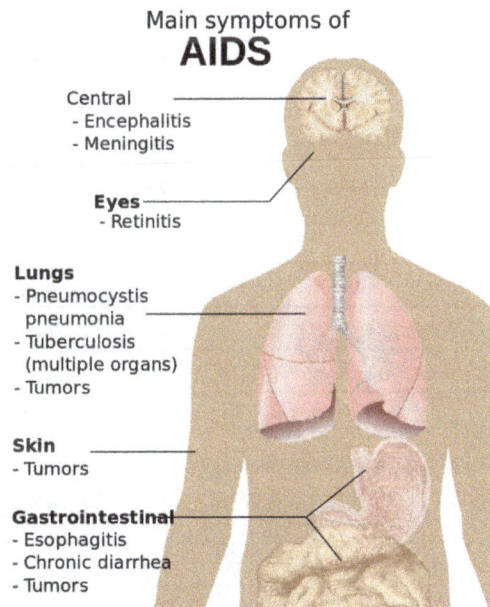

Main symptoms of AIDS.

Acquired immunodeficiency syndrome (AIDS) is defined in terms of either a CD4$^+$ T cell count below 200 cells per μL or the occurrence of specific diseases in association with an HIV infection. In the absence of specific treatment, around half of people infected with HIV develop AIDS within ten years. The most common initial conditions that alert to the presence of AIDS are pneumocystis pneumonia (40%), cachexia in the form of HIV wasting syndrome (20%), and esophageal candidiasis. Other common signs include recurring respiratory tract infections.

Opportunistic infections may be caused by bacteria, viruses, fungi, and parasites that are normally controlled by the immune system. Which infections occur depends partly on what organisms are common in the person's environment. These infections may affect nearly every organ system.

People with AIDS have an increased risk of developing various viral-induced cancers, including Kaposi's sarcoma, Burkitt's lymphoma, primary central nervous system lymphoma, and cervical cancer. Kaposi's sarcoma is the most common cancer occurring in 10 to 20% of people with HIV. The second most common cancer is lymphoma, which is the cause of death of nearly 16% of people with AIDS and is the initial sign of AIDS in 3 to 4%. Both these cancers are associated with human herpesvirus 8. Cervical cancer occurs more frequently in those with AIDS because of its association with human papillomavirus (HPV). Conjunctival cancer (of the layer that lines the inner part of eyelids and the white part of the eye) is also more common in those with HIV.

Additionally, people with AIDS frequently have systemic symptoms such as prolonged fevers, sweats (particularly at night), swollen lymph nodes, chills, weakness, and unintended weight loss. Diarrhea is another common symptom, present in about 90% of people with AIDS. They can also be affected by diverse psychiatric and neurological symptoms independent of opportunistic infections and cancers.

Transmission

Average per act risk of getting HIV by exposure route to an infected source	
Exposure route	Chance of infection
Blood transfusion	90%
Childbirth (to child)	25%
Needle-sharing injection drug use	0.67%
Percutaneous needle stick	0.30%
Receptive anal intercourse*	0.04–3.0%
Insertive anal intercourse*	0.03%
Receptive penile-vaginal intercourse*	0.05–0.30%
Insertive penile-vaginal intercourse*	0.01–0.38%
Receptive oral intercourse*§	0–0.04%
Insertive oral intercourse*§	0–0.005%
* assuming no condom use § source refers to oral intercourse performed on a man	

HIV is transmitted by three main routes: sexual contact, significant exposure to infected body fluids or tissues, and from mother to child during pregnancy, delivery, or breastfeeding (known as vertical transmission). There is no risk of acquiring HIV if exposed to feces, nasal secretions, saliva, sputum, sweat, tears, urine, or vomit unless these are contaminated with blood. It is possible to be co-infected by more than one strain of HIV—a condition known as HIV superinfection.

Sexual

The most frequent mode of transmission of HIV is through sexual contact with an infected person. The majority of all transmissions worldwide occur through heterosexual contacts (i.e. sexual contacts between people of the opposite sex); however, the pattern of transmission varies significantly among countries. In the United States, as of 2010, most transmission occurred in men who had sex with men, with this population accounting for 65% of all new cases.

With regard to unprotected heterosexual contacts, estimates of the risk of HIV transmission per sexual act appear to be four to ten times higher in low-income countries than in high-income countries. In low-income countries, the risk of female-to-male transmission is estimated as 0.38% per act, and of male-to-female transmission as 0.30% per act; the equivalent estimates for high-income countries are 0.04% per act for female-to-male transmission, and 0.08% per act for male-to-female transmission. The risk of transmission from anal intercourse is especially high, estimated as 1.4–1.7% per act in both heterosexual and homosexual contacts. While the risk of transmission from oral sex is relatively low, it is still present. The risk from receiving oral sex has been described as "nearly nil"; however, a few cases have been reported. The per-act risk is estimated at 0–0.04% for receptive oral intercourse. In settings involving prostitution in low income countries, risk of female-to-male transmission has been estimated as 2.4% per act and male-to-female transmission as 0.05% per act.

Risk of transmission increases in the presence of many sexually transmitted infections and genital ulcers. Genital ulcers appear to increase the risk approximately fivefold. Other sexually transmitted infections, such as gonorrhea, chlamydia, trichomoniasis, and bacterial vaginosis, are associated with somewhat smaller increases in risk of transmission.

The viral load of an infected person is an important risk factor in both sexual and mother-to-child transmission. During the first 2.5 months of an HIV infection a person's infectiousness is twelve times higher due to this high viral load. If the person is in the late stages of infection, rates of transmission are approximately eightfold greater.

Commercial sex workers (including those in pornography) have an increased rate of HIV. Rough sex can be a factor associated with an increased risk of transmission. Sexual assault is also believed to carry an increased risk of HIV transmission as condoms are rarely worn, physical trauma to the vagina or rectum is likely, and there may be a greater risk of concurrent sexually transmitted infections.

Body Fluids

The second most frequent mode of HIV transmission is via blood and blood products. Blood-borne transmission can be through needle-sharing during intravenous drug use, needle stick injury, transfusion of contaminated blood or blood product, or medical injections with unsterilised equipment. The risk from sharing a needle during drug injection is between 0.63 and 2.4% per act, with an average of 0.8%. The risk of acquiring HIV from a needle stick from an HIV-infected person is estimated as 0.3% (about 1 in 333) per act and the risk following mucous membrane exposure to infected blood as 0.09% (about 1 in 1000) per act. In the United States intravenous drug users made up 12% of all new cases of HIV in 2009, and in some areas more than 80% of people who inject drugs are HIV positive.

HIV is transmitted in about 93% of blood transfusions using infected blood. In developed countries the risk of acquiring HIV from a blood transfusion is extremely low (less than one in half a million) where improved donor selection and HIV screening is performed; for example, in the UK the risk is reported at one in five million and in the United States it was one in 1.5 million in 2008. In low income countries, only half of transfusions may be appropriately screened (as of 2008), and it is estimated that up to 15% of HIV infections in these areas come from transfusion of infected blood and blood products, representing between 5% and 10% of global infections. Although rare because of screening, it is possible to acquire HIV from organ and tissue transplantation.

CDC poster from 1989 highlighting the threat of AIDS associated with drug use

Unsafe medical injections play a significant role in HIV spread in sub-Saharan Africa. In 2007, between 12 and 17% of infections in this region were attributed to medical syringe use. The World Health Organization estimates the risk of transmission as a result of a medical injection in Africa at 1.2%. Significant risks are also associated with invasive procedures, assisted delivery, and dental care in this area of the world.

People giving or receiving tattoos, piercings, and scarification are theoretically at risk of infection but no confirmed cases have been documented. It is not possible for mosquitoes or other insects to transmit HIV.

Mother-To-Child

HIV can be transmitted from mother to child during pregnancy, during delivery, or through breast milk resulting in infection in the baby. This is the third most common way in which HIV is transmitted globally. In the absence of treatment, the risk of transmission before or during birth is around 20% and in those who also breastfeed 35%. As of 2008, vertical transmission accounted for about 90% of cases of HIV in children. With appropriate treatment the risk of mother-to-child infection can be reduced to about 1%. Preventive treatment involves the mother taking antiretrovirals during pregnancy and delivery, an elective caesarean section, avoiding breastfeeding, and administering antiretroviral drugs to the newborn. Antiretrovirals when taken by either the mother or the infant decrease the risk of transmission in those who do breastfeed. Many of these measures are however not available in the developing world. If blood contaminates food during pre-chewing it may pose a risk of transmission.

Virology

HIV is the cause of the spectrum of disease known as HIV/AIDS. HIV is a retrovirus that primarily infects components of the human immune system such as CD4$^+$ T cells, macrophages and dendritic cells. It directly and indirectly destroys CD4$^+$ T cells.

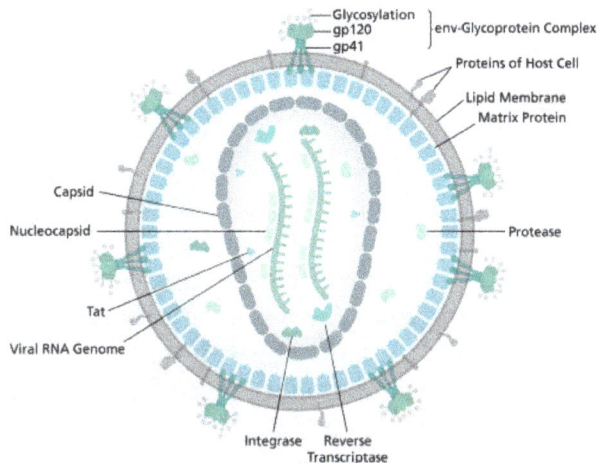

Diagram of a HIV virion structure

HIV is a member of the genus *Lentivirus*, part of the family *Retroviridae*. Lentiviruses share many morphological and biological characteristics. Many species of mammals are infected by lentiviruses, which are characteristically responsible for long-duration illnesses with a long incubation period. Lentiviruses are transmitted as single-stranded, positive-sense, enveloped RNA viruses. Upon entry into the target cell, the viral RNA genome is converted (reverse transcribed) into double-stranded DNA by a virally encoded reverse transcriptase that is transported along with the viral genome in the virus particle. The resulting viral DNA is then imported into the cell nucleus and integrated into the cellular DNA by a virally encoded integrase and host co-factors. Once integrated, the virus may become latent, allowing the virus and its host cell to avoid detection by the immune system. Alternatively, the virus may be transcribed, producing new RNA genomes and viral proteins that are packaged and released from the cell as new virus particles that begin the replication cycle anew.

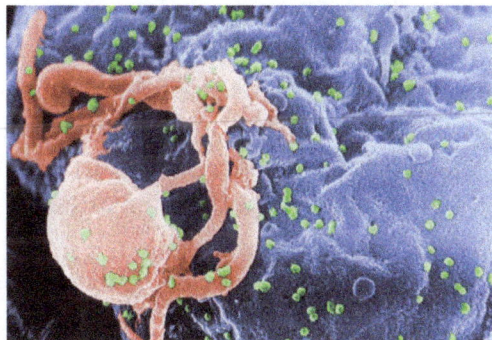

Scanning electron micrograph of HIV-1, colored green, budding from a cultured lymphocyte.

HIV is now known to spread between CD4+ T cells by two parallel routes: cell-free spread and cell-to-cell spread, i.e. it employs hybrid spreading mechanisms. In the cell-free spread, virus particles bud from an infected T cell, enter the blood/extracellular fluid and then infect another T cell following a chance encounter. HIV can also disseminate by direct transmission from one cell to another by a process of cell-to-cell spread. The hybrid spreading mechanisms of HIV contribute to the virus's ongoing replication against antiretroviral therapies.

Two types of HIV have been characterized: HIV-1 and HIV-2. HIV-1 is the virus that was originally discovered (and initially referred to also as LAV or HTLV-III). It is more virulent, more infective, and is the cause of the majority of HIV infections globally. The lower infectivity of HIV-2 as compared with HIV-1 implies that fewer people exposed to HIV-2 will be infected per exposure. Because of its relatively poor capacity for transmission, HIV-2 is largely confined to West Africa.

Pathophysiology

After the virus enters the body there is a period of rapid viral replication, leading to an abundance of virus in the peripheral blood. During primary infection, the level of HIV may reach several million virus particles per milliliter of blood. This response is accompanied by a marked drop in the number of circulating CD4+ T cells. The acute viremia is almost invariably associated with activation of CD8+ T cells, which kill HIV-infected cells, and subsequently with antibody production, or seroconversion. The CD8+ T cell response is thought to be important in controlling virus levels, which peak and then decline, as the CD4+ T cell counts recover. A good CD8+ T cell response has been linked to slower disease progression and a better prognosis, though it does not eliminate the virus.

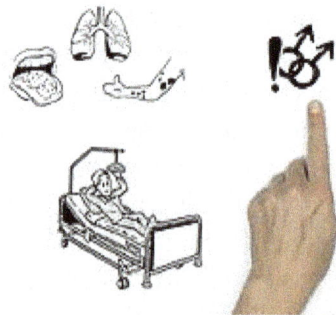

HIV/AIDS explained in a simple way

Ultimately, HIV causes AIDS by depleting CD4+ T cells. This weakens the immune system and allows opportunistic infections. T cells are essential to the immune response and without them, the body cannot fight infections or kill cancerous cells. The mechanism of CD4+ T cell depletion differs in the acute and chronic phases. During the acute phase, HIV-induced cell lysis and killing of infected cells by cytotoxic T cells accounts for CD4+ T cell depletion, although apoptosis may also be a factor. During the chronic phase, the consequences of generalized immune activation coupled with the gradual loss of the ability of the immune system to generate new T cells appear to account for the slow decline in CD4+ T cell numbers.

Although the symptoms of immune deficiency characteristic of AIDS do not appear for years after a person is infected, the bulk of CD4+ T cell loss occurs during the first weeks of infection, especially in the intestinal mucosa, which harbors the majority of the lymphocytes found in the body. The reason for the preferential loss of mucosal CD4+ T cells is that the majority of mucosal CD4+ T cells express the CCR5 protein which HIV uses as a co-receptor to gain access to the cells, whereas only a small fraction of CD4+ T cells in the bloodstream do so. A specific genetic change that alters the CCR5 protein when present in both chromosomes very effectively prevents HIV-1 infection.

HIV seeks out and destroys CCR5 expressing CD4+ T cells during acute infection. A vigorous immune response eventually controls the infection and initiates the clinically latent phase. CD4+ T cells in mucosal tissues remain particularly affected. Continuous HIV replication causes a state of generalized immune activation persisting throughout the chronic phase. Immune activation, which is reflected by the increased activation state of immune cells and release of pro-inflammatory cytokines, results from the activity of several HIV gene products and the immune response to ongoing HIV replication. It is also linked to the breakdown of the immune surveillance system of the gastrointestinal mucosal barrier caused by the depletion of mucosal CD4+ T cells during the acute phase of disease.

Diagnosis

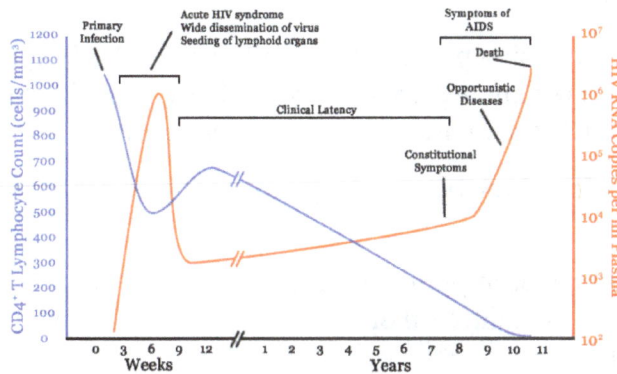

A generalized graph of the relationship between HIV copies (viral load) and CD4+ T cell counts over the average course of untreated HIV infection.

- CD4+ T Lymphocyte count (cells/mm^3)

- HIV RNA copies per mL of plasma

HIV/AIDS is diagnosed via laboratory testing and then staged based on the presence of certain signs or symptoms. HIV screening is recommended by the United States Preventive Services Task Force for all people 15 years to 65 years of age including all pregnant women. Additionally, testing is recommended for those at high risk, which includes anyone diagnosed with a sexually transmitted illness. In many areas of the world, a third of HIV carriers only discover they are infected at an advanced stage of the disease when AIDS or severe immunodeficiency has become apparent.

HIV Testing

Most people infected with HIV develop specific antibodies (i.e. seroconvert) within three to twelve weeks of the initial infection. Diagnosis of primary HIV before seroconversion is done by measuring HIV-RNA or p24 antigen. Positive results obtained by antibody or PCR testing are confirmed either by a different antibody or by PCR.

Antibody tests in children younger than 18 months are typically inaccurate due to the continued presence of maternal antibodies. Thus HIV infection can only be diagnosed by PCR testing for HIV RNA or DNA, or via testing for the p24 antigen. Much of the world lacks access to reliable PCR testing and many places simply wait until either symptoms develop or the child is old enough for accurate antibody testing. In sub-Saharan Africa as of 2007–2009 between 30 and 70% of the population were aware of their HIV status. In 2009, between 3.6 and 42% of men and women in Sub-Saharan countries were tested which represented a significant increase compared to previous years.

Classifications

Two main clinical staging systems are used to classify HIV and HIV-related disease for surveillance purposes: the WHO disease staging system for HIV infection and disease, and the CDC classification system for HIV infection. The CDC's classification system is more frequently adopted in developed countries. Since the WHO's staging system does not require laboratory tests, it is suited to the resource-restricted conditions encountered in developing countries, where it can also be used to help guide clinical management. Despite their differences, the two systems allow comparison for statistical purposes.

The World Health Organization first proposed a definition for AIDS in 1986. Since then, the WHO classification has been updated and expanded several times, with the most recent version being published in 2007. The WHO system uses the following categories:

- Primary HIV infection: May be either asymptomatic or associated with acute retroviral syndrome.

- Stage I: HIV infection is asymptomatic with a CD4$^+$ T cell count (also known as CD4 count) greater than 500 per microlitre (µl or cubic mm) of blood. May include generalized lymph node enlargement.

- Stage II: Mild symptoms which may include minor mucocutaneous manifestations and recurrent upper respiratory tract infections. A CD4 count of less than 500/µl.

- Stage III: Advanced symptoms which may include unexplained chronic diarrhea for longer than a month, severe bacterial infections including tuberculosis of the lung, and a CD4 count of less than 350/µl.

- Stage IV or AIDS: severe symptoms which include toxoplasmosis of the brain, candidiasis of the esophagus, trachea, bronchi or lungs and Kaposi's sarcoma. A CD4 count of less than 200/µl.

The United States Center for Disease Control and Prevention also created a classification system for HIV, and updated it in 2008 and 2014. This system classifies HIV infections based on CD4 count and clinical symptoms, and describes the infection in five groups. In those greater than six years of age it is:

- Stage 0: the time between a negative or indeterminate HIV test followed less than 180 days by a positive test

- Stage 1: CD4 count ≥ 500 cells/µl and no AIDS defining conditions

- Stage 2: CD4 count 200 to 500 cells/μl and no AIDS defining conditions

- Stage 3: CD4 count ≤ 200 cells/μl or AIDS defining conditions

- Unknown: if insufficient information is available to make any of the above classifications

For surveillance purposes, the AIDS diagnosis still stands even if, after treatment, the CD4+ T cell count rises to above 200 per μL of blood or other AIDS-defining illnesses are cured.

Prevention

AIDS Clinic, McLeod Ganj, Himachal Pradesh, India, 2010

Sexual Contact

Consistent condom use reduces the risk of HIV transmission by approximately 80% over the long term. When condoms are used consistently by a couple in which one person is infected, the rate of HIV infection is less than 1% per year. There is some evidence to suggest that female condoms may provide an equivalent level of protection. Application of a vaginal gel containing tenofovir (a reverse transcriptase inhibitor) immediately before sex seems to reduce infection rates by approximately 40% among African women. By contrast, use of the spermicide nonoxynol-9 may increase the risk of transmission due to its tendency to cause vaginal and rectal irritation.

Circumcision in Sub-Saharan Africa "reduces the acquisition of HIV by heterosexual men by between 38% and 66% over 24 months". Due to these studies, both the World Health Organization and UNAIDS recommended male circumcision as a method of preventing female-to-male HIV transmission in 2007 in areas with a high rates of HIV. However, whether it protects against male-to-female transmission is disputed, and whether it is of benefit in developed countries and among men who have sex with men is undetermined. The International Antiviral Society, however, does recommend for all sexually active heterosexual males and that it be discussed as an option with men who have sex with men. Some experts fear that a lower perception of vulnerability among circumcised men may cause more sexual risk-taking behavior, thus negating its preventive effects.

Programs encouraging sexual abstinence do not appear to affect subsequent HIV risk. Evidence of any benefit from peer education is equally poor. Comprehensive sexual education provided at school may decrease high risk behavior. A substantial minority of young people continues to engage in high-risk practices despite knowing about HIV/AIDS, underestimating their own risk of becoming infected with HIV. Voluntary counseling and testing people for HIV does not affect risky behavior in those who test negative but does increase condom use in those who test positive. It is not known whether treating other sexually transmitted infections is effective in preventing HIV.

Pre-Exposure

Antiretroviral treatment among people with HIV whose CD4 count ≤ 550 cells/μL is a very effective way to prevent HIV infection of their partner (a strategy known as treatment as prevention, or TASP). TASP is associated with a 10 to 20 fold reduction in transmission risk. Pre-exposure prophylaxis (PrEP) with a daily dose of the medications tenofovir, with or without emtricitabine, is effective in a number of groups including men who have sex with men, couples where one is HIV positive, and young heterosexuals in Africa. It may also be effective in intravenous drug users with a study finding a decrease in risk of 0.7 to 0.4 per 100 person years.

Universal precautions within the health care environment are believed to be effective in decreasing the risk of HIV. Intravenous drug use is an important risk factor and harm reduction strategies such as needle-exchange programs and opioid substitution therapy appear effective in decreasing this risk.

Post-Exposure

A course of antiretrovirals administered within 48 to 72 hours after exposure to HIV-positive blood or genital secretions is referred to as post-exposure prophylaxis (PEP). The use of the single agent zidovudine reduces the risk of a HIV infection five-fold following a needle-stick injury. As of 2013, the prevention regimen recommended in the United States consists of three medications—tenofovir, emtricitabine and raltegravir—as this may reduce the risk further.

PEP treatment is recommended after a sexual assault when the perpetrator is known to be HIV positive, but is controversial when their HIV status is unknown. The duration of treatment is usually four weeks and is frequently associated with adverse effects—where zidovudine is used, about 70% of cases result in adverse effects such as nausea (24%), fatigue (22%), emotional distress (13%) and headaches (9%).

Mother-To-Child

Programs to prevent the vertical transmission of HIV (from mothers to children) can reduce rates of transmission by 92–99%. This primarily involves the use of a combination of antiviral medications during pregnancy and after birth in the infant and potentially includes bottle feeding rather than breastfeeding. If replacement feeding is acceptable, feasible, affordable, sustainable, and safe, mothers should avoid breastfeeding their infants; however exclusive breastfeeding is recommended during the first months of life if this is not the case. If exclusive breastfeeding is carried out, the provision of extended antiretroviral prophylaxis to the infant decreases the risk of transmission. In 2015, Cuba became the first country in the world to eradicate mother-to-child transmission of HIV.

Vaccination

Currently, there is no licensed vaccine for HIV or AIDS. The most effective vaccine trial to date, RV 144, was published in 2009 and found a partial reduction in the risk of transmission of roughly 30%, stimulating some hope in the research community of developing a truly effective vaccine. Further trials of the RV 144 vaccine are ongoing.

Treatment

There is currently no cure or effective HIV vaccine. Treatment consists of highly active antiretroviral therapy (HAART) which slows progression of the disease. As of 2010 more than 6.6 million people were taking them in low and middle income countries. Treatment also includes preventive and active treatment of opportunistic infections.

Antiviral Therapy

Current HAART options are combinations (or "cocktails") consisting of at least three medications belonging to at least two types, or "classes," of antiretroviral agents. Initially treatment is typically a non-nucleoside reverse transcriptase inhibitor (NNRTI) plus two nucleoside analogue reverse transcriptase inhibitors (NRTIs). Typical NRTIs include: zidovudine (AZT) or tenofovir (TDF) and lamivudine (3TC) or emtricitabine (FTC). Combinations of agents which include protease inhibitors (PI) are used if the above regimen loses effectiveness.

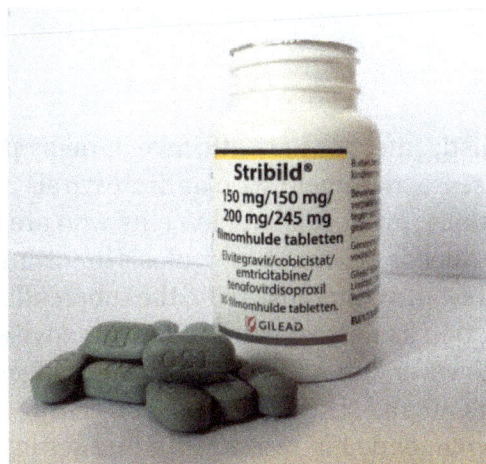

Stribild – a common once-daily ART regime consisting of elvitegravir, emtricitabine, tenofovir and the booster cobicistat

The World Health Organization and United States recommends antiretrovirals in people of all ages including pregnant women as soon as the diagnosis is made regardless of CD4 count. Once treatment is begun it is recommended that it is continued without breaks or "holidays". Many people are diagnosed only after treatment ideally should have begun. The desired outcome of treatment is a long term plasma HIV-RNA count below 50 copies/mL. Levels to determine if treatment is effective are initially recommended after four weeks and once levels fall below 50 copies/mL checks every three to six months are typically adequate. Inadequate control is deemed to be greater than 400 copies/mL. Based on these criteria treatment is effective in more than 95% of people during the first year.

Benefits of treatment include a decreased risk of progression to AIDS and a decreased risk of death. In the developing world treatment also improves physical and mental health. With treatment there is a 70% reduced risk of acquiring tuberculosis. Additional benefits include a decreased risk of transmission of the disease to sexual partners and a decrease in mother-to-child transmission. The effectiveness of treatment depends to a large part on compliance. Reasons for non-adherence include poor access to medical care, inadequate social supports, mental illness and drug abuse. The complexity of treatment regimens (due to pill numbers and dosing frequency) and adverse effects may reduce adherence. Even though cost is an important issue with some medications, 47% of those who needed them were taking them in low and middle income countries as of 2010 and the rate of adherence is similar in low-income and high-income countries.

Specific adverse events are related to the antiretroviral agent taken. Some relatively common adverse events include: lipodystrophy syndrome, dyslipidemia, and diabetes mellitus, especially with protease inhibitors. Other common symptoms include diarrhea, and an increased risk of cardiovascular disease. Newer recommended treatments are associated with fewer adverse effects. Certain medications may be associated with birth defects and therefore may be unsuitable for women hoping to have children.

Treatment recommendations for children are somewhat different from those for adults. The World Health Organisation recommends treating all children less than 5 years of age; children above 5 are treated like adults. The United States guidelines recommend treating all children less than 12 months of age and all those with HIV RNA counts greater than 100,000 copies/mL between one year and five years of age.

Opportunistic Infections

Measures to prevent opportunistic infections are effective in many people with HIV/AIDS. In addition to improving current disease, treatment with antiretrovirals reduces the risk of developing additional opportunistic infections. Adults and adolescents who are living with HIV (even on anti-retroviral therapy) with no evidence of active tuberculosis in settings with high tuberculosis burden should receive isoniazid preventive therapy (IPT), the tuberculin skin test can be used to help decide if IPT is needed. Vaccination against hepatitis A and B is advised for all people at risk of HIV before they become infected; however it may also be given after infection. Trimethoprim/sulfamethoxazole prophylaxis between four and six weeks of age and ceasing breastfeeding in infants born to HIV positive mothers is recommended in resource limited settings. It is also recommended to prevent PCP when a person's CD4 count is below 200 cells/uL and in those who have or have previously had PCP. People with substantial immunosuppression are also advised to receive prophylactic therapy for toxoplasmosis and Cryptococcus meningitis. Appropriate preventive measures have reduced the rate of these infections by 50% between 1992 and 1997.

Diet

The World Health Organization (WHO) has issued recommendations regarding nutrient requirements in HIV/AIDS. A generally healthy diet is promoted. Some evidence has shown a benefit from micronutrient supplements. Evidence for supplementation with selenium is mixed with some tentative evidence of benefit. There is some evidence that vitamin A supplementation in children reduces mortality and improves growth. In Africa in nutritionally compromised pregnant and lactating women a multivitamin supplementation has improved outcomes for both mothers and children. Dietary intake of micronutrients at RDA levels by HIV-infected adults is recommended by the WHO; higher intake of vitamin A, zinc, and iron can produce adverse effects in HIV positive adults, and is not recommended unless there is documented deficiency.

Alternative Medicine

In the US, approximately 60% of people with HIV use various forms of complementary or alternative medicine, even though the effectiveness of most of these therapies has not been established. There is not enough evidence to support the use of herbal medicines. There is insufficient evidence to recommend or support the use of medical cannabis to try to increase appetite or weight gain.

Prognosis

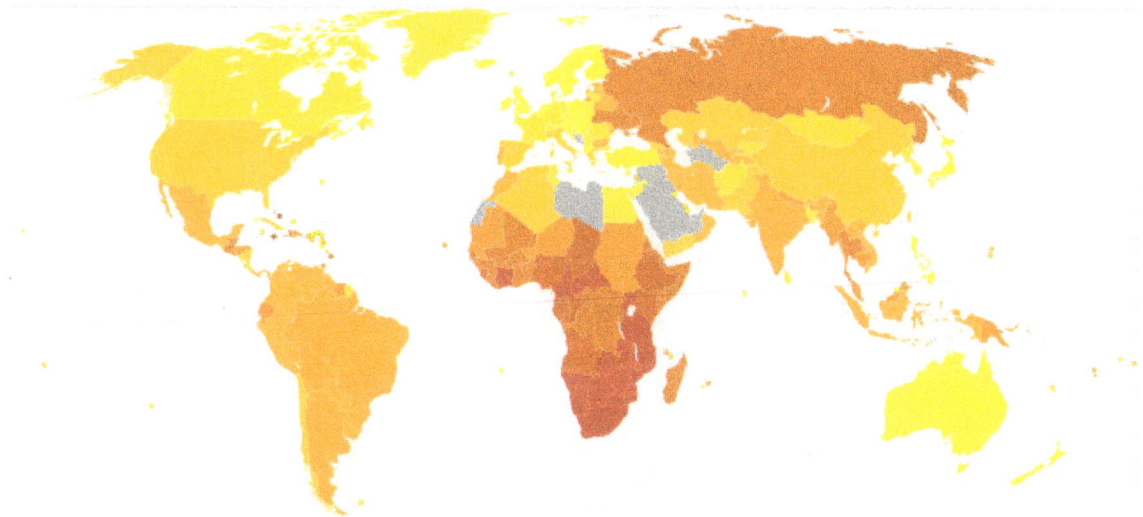

Deaths Due To Hiv/Aids Per Million Persons In 2012

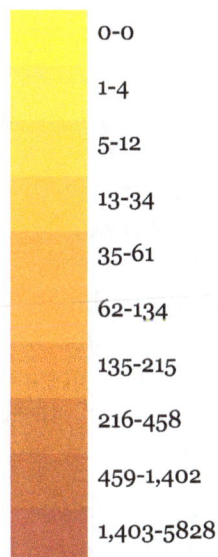

0-0

1-4

5-12

13-34

35-61

62-134

135-215

216-458

459-1,402

1,403-5828

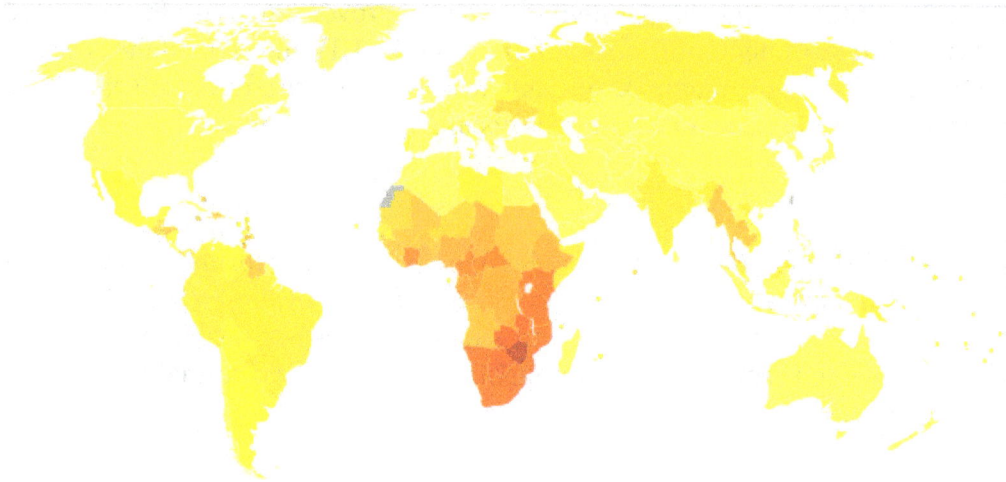

Disability-adjusted life year for HIV and AIDS per 100,000 inhabitants as of 2004.

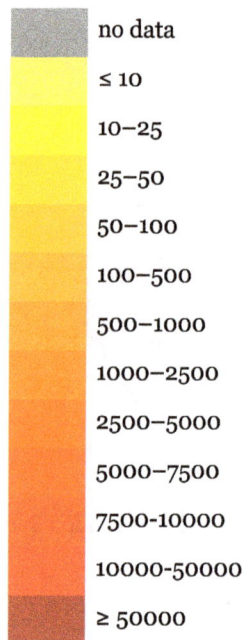

	no data
	≤ 10
	10–25
	25–50
	50–100
	100–500
	500–1000
	1000–2500
	2500–5000
	5000–7500
	7500-10000
	10000-50000
	≥ 50000

HIV/AIDS has become a chronic rather than an acutely fatal disease in many areas of the world. Prognosis varies between people, and both the CD4 count and viral load are useful for predicted outcomes. Without treatment, average survival time after infection with HIV is estimated to be 9 to 11 years, depending on the HIV subtype. After the diagnosis of AIDS, if treatment is not available, survival ranges between 6 and 19 months. HAART and appropriate prevention of opportunistic infections reduces the death rate by 80%, and raises the life expectancy for a newly diagnosed young adult to 20–50 years. This is between two thirds and nearly that of the general population. If treatment is started late in the infection, prognosis is not as good: for example, if treatment is begun following the diagnosis of AIDS, life expectancy is ~10–40 years. Half of infants born with HIV die before two years of age without treatment.

The primary causes of death from HIV/AIDS are opportunistic infections and cancer, both of which are frequently the result of the progressive failure of the immune system. Risk of cancer appears to increase once the CD4 count is below 500/µL. The rate of clinical disease progression varies widely between individuals and has been shown to be affected by a number of factors such as a person's susceptibility and immune function; their access to health care, the presence of co-infections; and the particular strain (or strains) of the virus involved.

Tuberculosis co-infection is one of the leading causes of sickness and death in those with HIV/AIDS being present in a third of all HIV-infected people and causing 25% of HIV-related deaths. HIV is also one of the most important risk factors for tuberculosis. Hepatitis C is another very common co-infection where each disease increases the progression of the other. The two most common cancers associated with HIV/AIDS are Kaposi's sarcoma and AIDS-related non-Hodgkin's lymphoma.

Even with anti-retroviral treatment, over the long term HIV-infected people may experience neurocognitive disorders, osteoporosis, neuropathy, cancers, nephropathy, and cardiovascular disease. Some conditions like lipodystrophy may be caused both by HIV and its treatment.

Epidemiology

HIV/AIDS is a global pandemic. As of 2014, approximately 37 million people have HIV worldwide with the number of new infections that year being about 2 million. This is down from 3.1 million new infections in 2001. Of these 37 million more than half are women and 2.6 million are less than 15 years old. It resulted in about 1.2 million deaths in 2014, down from a peak of 2.2 million in 2005.

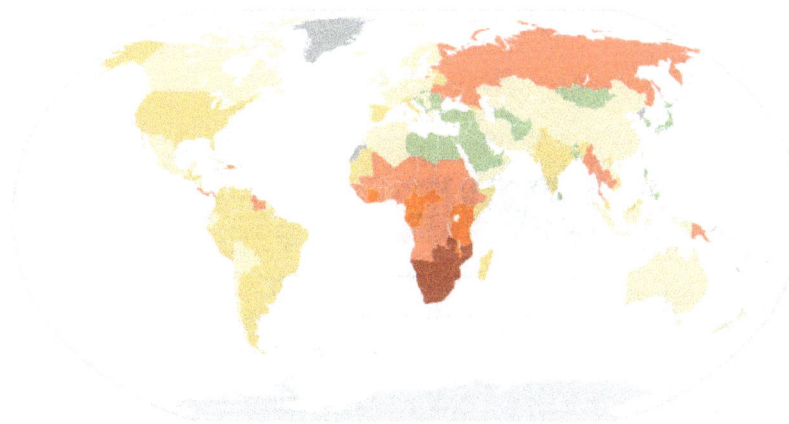

Estimated percentage of HIV among young adults (15–49) per country as of 2011.

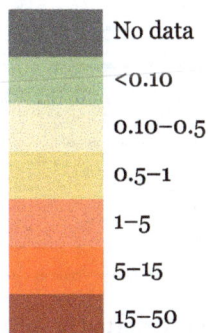

■	No data
■	<0.10
■	0.10–0.5
■	0.5–1
■	1–5
■	5–15
■	15–50

Sub-Saharan Africa is the region most affected. In 2010, an estimated 68% (22.9 million) of all HIV cases and 66% of all deaths (1.2 million) occurred in this region. This means that about 5% of the adult population is infected and it is believed to be the cause of 10% of all deaths in children. Here in contrast to other regions women compose nearly 60% of cases. South Africa has the largest population of people with HIV of any country in the world at 5.9 million. Life expectancy has fallen in the worst-affected countries due to HIV/AIDS; for example, in 2006 it was estimated that it had dropped from 65 to 35 years in Botswana. Mother-to-child transmission, as of 2013, in Botswana and South Africa has decreased to less than 5% with improvement in many other African nations due to improved access to antiretroviral therapy.

South & South East Asia is the second most affected; in 2010 this region contained an estimated 4 million cases or 12% of all people living with HIV resulting in approximately 250,000 deaths. Approximately 2.4 million of these cases are in India.

In 2008 in the United States approximately 1.2 million people were living with HIV, resulting in about 17,500 deaths. The US Centers for Disease Control and Prevention estimated that in 2008 20% of infected Americans were unaware of their infection. In the United Kingdom as of 2009 there were approximately 86,500 cases which resulted in 516 deaths. In Canada as of 2008 there were about 65,000 cases causing 53 deaths. Between the first recognition of AIDS in 1981 and 2009 it has led to nearly 30 million deaths. Prevalence is lowest in Middle East and North Africa at 0.1% or less, East Asia at 0.1% and Western and Central Europe at 0.2%. The worst affected European countries, in 2009 and 2012 estimates, are Russia, Ukraine, Latvia, Moldova, Portugal and Belarus, in decreasing order of prevalence.

History

Discovery

The *Morbidity and Mortality Weekly Report* reported in 1981 on what was later to be called "AIDS".

AIDS was first clinically observed in 1981 in the United States. The initial cases were a cluster of injecting drug users and homosexual men with no known cause of impaired immunity who showed symptoms of *Pneumocystis carinii* pneumonia (PCP), a rare opportunistic infection that was known to occur in people with very compromised immune systems. Soon thereafter, an unexpected number of homosexual men developed a previously rare skin cancer called Kaposi's sarcoma (KS). Many more cases of PCP and KS emerged, alerting U.S. Centers for Disease Control and Prevention (CDC) and a CDC task force was formed to monitor the outbreak.

In the early days, the CDC did not have an official name for the disease, often referring to it by way of the diseases that were associated with it, for example, lymphadenopathy, the disease after which the discoverers of HIV originally named the virus. They also used *Kaposi's sarcoma and opportunistic infections*, the name by which a task force had been set up in 1981. At one point, the CDC coined the phrase "the 4H disease", since the syndrome seemed to affect heroin users, homosexuals, hemophiliacs, and Haitians. In the general press, the term "GRID", which stood for gay-related immune deficiency, had been coined. However, after determining that AIDS was not isolated to the gay community, it was realized that the term GRID was misleading and the term AIDS was introduced at a meeting in July 1982. By September 1982 the CDC started referring to the disease as AIDS.

In 1983, two separate research groups led by Robert Gallo and Luc Montagnier declared that a novel retrovirus may have been infecting people with AIDS, and published their findings in the same issue of the journal *Science*. Gallo claimed that a virus his group had isolated from a person with AIDS was strikingly similar in shape to other human T-lymphotropic viruses (HTLVs) his group had been the first to isolate. Gallo's group called their newly isolated virus HTLV-III. At the same time, Montagnier's group isolated a virus from a person presenting with swelling of the lymph nodes of the neck and physical weakness, two characteristic symptoms of AIDS. Contradicting the report from Gallo's group, Montagnier and his colleagues showed that core proteins of this virus were immunologically different from those of HTLV-I. Montagnier's group named their isolated virus lymphadenopathy-associated virus (LAV). As these two viruses turned out to be the same, in 1986, LAV and HTLV-III were renamed HIV.

Origins

Both HIV-1 and HIV-2 are believed to have originated in non-human primates in West-central Africa and were transferred to humans in the early 20th century. HIV-1 appears to have originated in southern Cameroon through the evolution of SIV(cpz), a simian immunodeficiency virus (SIV) that infects wild chimpanzees (HIV-1 descends from the SIVcpz endemic in the chimpanzee subspecies *Pan troglodytes troglodytes*). The closest relative of HIV-2 is SIV(smm), a virus of the sooty mangabey (*Cercocebus atys atys*), an Old World monkey living in coastal West Africa (from southern Senegal to western Côte d'Ivoire). New World monkeys such as the owl monkey are resistant to HIV-1 infection, possibly because of a genomic fusion of two viral resistance genes. HIV-1 is thought to have jumped the species barrier on at least three separate occasions, giving rise to the three groups of the virus, M, N, and O.

There is evidence that humans who participate in bushmeat activities, either as hunters or as bushmeat vendors, commonly acquire SIV. However, SIV is a weak virus which is typically suppressed by the human immune system within weeks of infection. It is thought that several transmissions of the virus from individual to individual in quick succession are necessary to allow it enough time to mutate into HIV. Furthermore, due to its relatively low person-to-person transmission rate, SIV can only spread throughout the population in the presence of one or more high-risk transmission channels, which are thought to have been absent in Africa before the 20th century.

Left to right: the African green monkey source of SIV, the sooty mangabey source of HIV-2 and the chimpanzee source of HIV-1

Specific proposed high-risk transmission channels, allowing the virus to adapt to humans and spread throughout the society, depend on the proposed timing of the animal-to-human crossing. Genetic studies of the virus suggest that the most recent common ancestor of the HIV-1 M group dates back to circa 1910. Proponents of this dating link the HIV epidemic with the emergence of colonialism and growth of large colonial African cities, leading to social changes, including a higher degree of sexual promiscuity, the spread of prostitution, and the accompanying high frequency of genital ulcer diseases (such as syphilis) in nascent colonial cities. While transmission rates of HIV during vaginal intercourse are low under regular circumstances, they are increased many fold if one of the partners suffers from a sexually transmitted infection causing genital ulcers. Early 1900s colonial cities were notable due to their high prevalence of prostitution and genital ulcers, to the degree that, as of 1928, as many as 45% of female residents of eastern Kinshasa were thought to have been prostitutes, and, as of 1933, around 15% of all residents of the same city had syphilis.

An alternative view holds that unsafe medical practices in Africa after World War II, such as unsterile reuse of single use syringes during mass vaccination, antibiotic and anti-malaria treatment campaigns, were the initial vector that allowed the virus to adapt to humans and spread.

The earliest well-documented case of HIV in a human dates back to 1959 in the Congo. The earliest retrospectively described case of AIDS is believed to have been in Norway beginning in 1966. In July 1960, in the wake its independence, the United Nations recruited Francophone experts and technicians from all over the world to assist in filling administrative gaps left by Belgium, who did not leave behind an African elite to run the country. By 1962, Haitians made up the second largest group of well-educated experts (out of the 48 national groups recruited), that totaled around 4500 in the country. Dr. Jacques Pépin, a Quebecer author of *The Origins of AIDS*, stipulates that Haiti was one of HIV's entry points to the United States and that one of them may have carried HIV back across the Atlantic in the 1960s. Although, the virus may have been present in the United States as early as 1966, the vast majority of infections occurring outside sub-Saharan Africa (including the U.S.) can be traced back to a single unknown individual who became infected with HIV in Haiti and then brought the infection to the United States some time around 1969. The epidemic then rapidly spread among high-risk groups (initially, sexually promiscuous men who have sex with men). By 1978, the prevalence of HIV-1 among homosexual male residents of New York and San Francisco was estimated at 5%, suggesting that several thousand individuals in the country had been infected.

Society and Culture

Stigma

AIDS stigma exists around the world in a variety of ways, including ostracism, rejection, discrimination and avoidance of HIV infected people; compulsory HIV testing without prior consent or protection of confidentiality; violence against HIV infected individuals or people who are perceived to be infected with HIV; and the quarantine of HIV infected individuals. Stigma-related violence or the fear of violence prevents many people from seeking HIV testing, returning for their results, or securing treatment, possibly turning what could be a manageable chronic illness into a death sentence and perpetuating the spread of HIV.

Ryan White became a poster child for HIV after being expelled from school because he was infected.

AIDS stigma has been further divided into the following three categories:

- *Instrumental AIDS stigma*—a reflection of the fear and apprehension that are likely to be associated with any deadly and transmissible illness.

- *Symbolic AIDS stigma*—the use of HIV/AIDS to express attitudes toward the social groups or lifestyles perceived to be associated with the disease.

- *Courtesy AIDS stigma*—stigmatization of people connected to the issue of HIV/AIDS or HIV-positive people.

Often, AIDS stigma is expressed in conjunction with one or more other stigmas, particularly those associated with homosexuality, bisexuality, promiscuity, prostitution, and intravenous drug use.

In many developed countries, there is an association between AIDS and homosexuality or bisexuality, and this association is correlated with higher levels of sexual prejudice, such as anti-homosexual/bisexual attitudes. There is also a perceived association between AIDS and all male-male sexual behavior, including sex between uninfected men. However, the dominant mode of spread worldwide for HIV remains heterosexual transmission.

In 2003, as part of an overall reform of marriage and population legislation, it became legal for people with AIDS to marry in China.

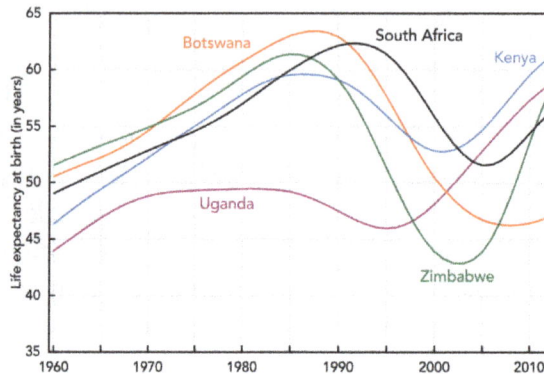

Changes in life expectancy in some African countries, 1960–2012

HIV/AIDS affects the economics of both individuals and countries. The gross domestic product of the most affected countries has decreased due to the lack of human capital. Without proper nutrition, health care and medicine, large numbers of people die from AIDS-related complications. They will not only be unable to work, but will also require significant medical care. It is estimated that as of 2007 there were 12 million AIDS orphans. Many are cared for by elderly grandparents.

Returning to work after beginning treatment for HIV/AIDS is difficult, and affected people often work less than the average worker. Unemployment in people with HIV/AIDS also is associated with suicidal ideation, memory problems, and social isolation; employment increases self-esteem, sense of dignity, confidence, and quality of life. A 2015 Cochrane review found low-quality evidence that antiretroviral treatment helps people with HIV/AIDS work more, and increases the chance that a person with HIV/AIDS will be employed.

By affecting mainly young adults, AIDS reduces the taxable population, in turn reducing the resources available for public expenditures such as education and health services not related to AIDS resulting in increasing pressure for the state's finances and slower growth of the economy. This causes a slower growth of the tax base, an effect that is reinforced if there are growing expenditures on treating the sick, training (to replace sick workers), sick pay and caring for AIDS orphans. This is especially true if the sharp increase in adult mortality shifts the responsibility and blame from the family to the government in caring for these orphans.

At the household level, AIDS causes both loss of income and increased spending on healthcare. A study in Côte d'Ivoire showed that households having a person with HIV/AIDS spent twice as much on medical expenses as other households. This additional expenditure also leaves less income to spend on education and other personal or family investment.

Religion and AIDS

The topic of religion and AIDS has become highly controversial in the past twenty years, primarily because some religious authorities have publicly declared their opposition to the use of condoms.

The religious approach to prevent the spread of AIDS according to a report by American health expert Matthew Hanley titled *The Catholic Church and the Global AIDS Crisis* argues that cultural changes are needed including a re-emphasis on fidelity within marriage and sexual abstinence outside of it.

Some religious organisations have claimed that prayer can cure HIV/AIDS. In 2011, the BBC reported that some churches in London were claiming that prayer would cure AIDS, and the Hackney-based Centre for the Study of Sexual Health and HIV reported that several people stopped taking their medication, sometimes on the direct advice of their pastor, leading to a number of deaths. The Synagogue Church Of All Nations advertise an "anointing water" to promote God's healing, although the group deny advising people to stop taking medication.

Media Portrayal

One of the first high-profile cases of AIDS was the American Rock Hudson, a gay actor who had been married and divorced earlier in life, who died on October 2, 1985 having announced that he was suffering from the virus on July 25 that year. He had been diagnosed during 1984. A notable British casualty of AIDS that year was Nicholas Eden, a gay politician and son of the late prime minister Anthony Eden. On November 24, 1991, the virus claimed the life of British rock star Freddie Mercury, lead singer of the band Queen, who died from an AIDS-related illness having only revealed the diagnosis on the previous day. However, he had been diagnosed as HIV positive in 1987. One of the first high-profile heterosexual cases of the virus was Arthur Ashe, the American tennis player. He was diagnosed as HIV positive on August 31, 1988, having contracted the virus from blood transfusions during heart surgery earlier in the 1980s. Further tests within 24 hours of the initial diagnosis revealed that Ashe had AIDS, but he did not tell the public about his diagnosis until April 1992. He died as a result on February 6, 1993 at age 49.

Therese Frare's photograph of gay activist David Kirby, as he lay dying from AIDS while surrounded by family, was taken in April 1990. *LIFE magazine* said the photo became the one image "most powerfully identified with the HIV/AIDS epidemic." The photo was displayed in *LIFE magazine*, was the winner of the World Press Photo, and acquired worldwide notoriety after being used in a United Colors of Benetton advertising campaign in 1992. In 1996, Johnson Aziga, a Ugandan-born Canadian was diagnosed with HIV, but subsequently had unprotected sex with 11 women without disclosing his diagnosis. By 2003 seven had contracted HIV, and two died from complications related to AIDS. Aziga was convicted of first-degree murder and is liable to a life sentence.

Criminal Transmission

Criminal transmission of HIV is the intentional or reckless infection of a person with the human immunodeficiency virus (HIV). Some countries or jurisdictions, including some areas of the United States, have laws that criminalize HIV transmission or exposure. Others may charge the accused under laws enacted before the HIV pandemic.

Misconceptions

There are many misconceptions about HIV and AIDS. Three of the most common are that AIDS can spread through casual contact, that sexual intercourse with a virgin will cure AIDS, and that

HIV can infect only gay men and drug users. In 2014, some among the British public wrongly thought you could get HIV from kissing (16%), sharing a glass (5%), spitting (16%), a public toilet seat (4%), and coughing or sneezing (5%). Other misconceptions are that any act of anal intercourse between two uninfected gay men can lead to HIV infection, and that open discussion of HIV and homosexuality in schools will lead to increased rates of AIDS.

A small group of individuals continue to dispute the connection between HIV and AIDS, the existence of HIV itself, or the validity of HIV testing and treatment methods. These claims, known as AIDS denialism, have been examined and rejected by the scientific community. However, they have had a significant political impact, particularly in South Africa, where the government's official embrace of AIDS denialism (1999–2005) was responsible for its ineffective response to that country's AIDS epidemic, and has been blamed for hundreds of thousands of avoidable deaths and HIV infections.

Several discredited conspiracy theories have held that HIV was created by scientists, either inadvertently or deliberately. Operation INFEKTION was a worldwide Soviet active measures operation to spread the claim that the United States had created HIV/AIDS. Surveys show that a significant number of people believed – and continue to believe – in such claims.

Research

HIV/AIDS research includes all medical research which attempts to prevent, treat, or cure HIV/AIDS along with fundamental research about the nature of HIV as an infectious agent and AIDS as the disease caused by HIV.

Many governments and research institutions participate in HIV/AIDS research. This research includes behavioral health interventions such as sex education, and drug development, such as research into microbicides for sexually transmitted diseases, HIV vaccines, and antiretroviral drugs. Other medical research areas include the topics of pre-exposure prophylaxis, post-exposure prophylaxis, and circumcision and HIV.

Chronic Granulomatous Disease

Chronic granulomatous disease (CGD) (also known as Bridges–Good syndrome, Chronic granulomatous disorder, and Quie syndrome) is a diverse group of hereditary diseases in which certain cells of the immune system have difficulty forming the reactive oxygen compounds (most importantly the superoxide radical due to defective phagocyte NADPH oxidase) used to kill certain ingested pathogens. This leads to the formation of granulomata in many organs. CGD affects about 1 in 200,000 people in the United States, with about 20 new cases diagnosed each year.

This condition was first discovered in 1950 in a series of 4 boys from Minnesota, and in 1957 was named "a fatal granulomatosus of childhood" in a publication describing their disease. The underlying cellular mechanism that causes chronic granulomatous disease was discovered in 1967, and research since that time has further elucidated the molecular mechanisms underlying the disease. Bernard Babior made key contributions in linking the defect of superoxide production of white blood cells, to the etiology of the disease. In 1986, the X-linked form of CGD was the first disease for which positional cloning was used to identify the underlying genetic mutation.

Classification

Chronic granulomatous disease is the name for a genetically heterogeneous group of immunodeficiencies. The core defect is a failure of phagocytic cells to kill organisms that they have engulfed because of defects in a system of enzymes that produce free radicals and other toxic small molecules. There are several types, including:

- X-linked chronic granulomatous disease]] (CGD)

- autosomal recessive cytochrome b-negative CGD

- autosomal recessive cytochrome b-positive CGD type I

- autosomal recessive cytochrome b-positive CGD type II

- atypical granulomatous disease

Symptoms

Classically, patients with chronic granulomatous disease will suffer from recurrent bouts of infection due to the decreased capacity of their immune system to fight off disease-causing organisms. The recurrent infections they acquire are specific and are, in decreasing order of frequency:

- pneumonia

- abscesses of the skin, tissues, and organs

- suppurative arthritis

- osteomyelitis

- bacteremia/fungemia

- superficial skin infections such as cellulitis or impetigo

Most people with CGD are diagnosed in childhood, usually before age 5. Early diagnosis is important since these people can be placed on antibiotics to ward off infections before they occur. Small groups of CGD patients may also be affected by McLeod syndrome because of the proximity of the two genes on the same X-chromosome.

Atypical Infections

People with CGD are sometimes infected with organisms that usually do not cause disease in people with normal immune systems. Among the most common organisms that cause disease in CGD patients are:

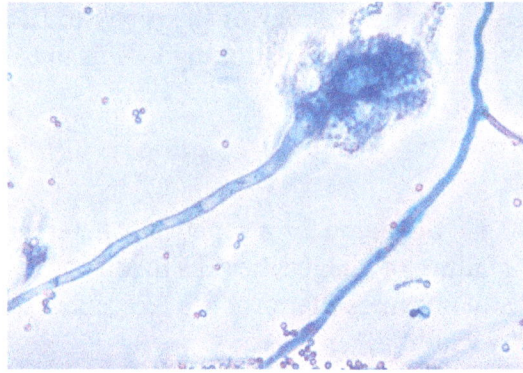

Microscopic image of the fungus, *Aspergillus fumigatus*, an organism that commonly causes disease in people with chronic granulomatous disease.

- bacteria (particularly those that are catalase-positive)

    o *Staphylococcus aureus*.

    o *Serratia marcescens*.

    o *Listeria* species.

    o *E. coli*.

    o *Klebsiella* species.

    o *Pseudomonas cepacia, a.k.a. Burkholderia cepacia*.

    o *Nocardia*.

- fungi

    o *Aspergillus* species. Aspergillus has a propensity to cause infection in people with CGD and of the Aspergillus species, *Aspergillus fumigatus* seems to be most common in CGD.

    o *Candida* species.

Patients with CGD can usually resist infections of catalase-negative bacteria. Catalase is an enzyme that catalyzes the breakdown of hydrogen peroxide in many organisms. In organisms that lack catalase (catalase-negative), normal metabolic functions will cause an accumulation of hydrogen peroxide which the host's immune system can use to fight off the infection. In organisms that have catalase (catalase-positive), the enzyme breaks down any hydrogen peroxide that was produced through normal metabolism. Therefore, hydrogen peroxide will not accumulate, leaving the patient vulnerable to catalase-positive bacteria.

Genetics

Most cases of chronic granulomatous disease are transmitted as a mutation on the X chromosome and are thus called an "X-linked trait". The affected gene on the X chromosome codes for the gp91 protein p91-PHOX (*p* is the weight of the protein in kDa; the *g* means glycoprotein). CGD can also be transmitted in an autosomal recessive fashion (via CYBA and NCF1) and affects other PHOX proteins. The type of mutation that causes both types of CGD are varied and may be deletions, frame-shift, nonsense, and missense.

A low level of NADPH, the cofactor required for superoxide synthesis, can lead to CGD. This has been reported in women who are homozygous for the genetic defect causing glucose-6-phosphate dehydrogenase deficiency (G6PD), which is characterised by reduced NADPH levels.

Pathophysiology

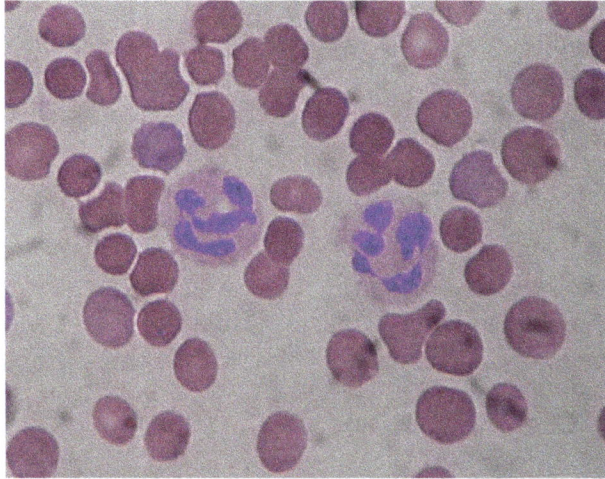

Two neutrophils among many red blood cells. Neutrophils are one type of cell affected by chronic granulomatous disease.

Phagocytes (i.e., neutrophils and macrophages) require an enzyme to produce reactive oxygen species to destroy bacteria after they are ingested (phagocytosis), a process known as the respiratory burst. This enzyme is termed "phagocyte NADPH oxidase" (*PHOX*). This enzyme oxidizes NADPH and reduces molecular oxygen to produce superoxide anions, a reactive oxygen species. Superoxide is then disproportionated into peroxide and molecular oxygen by superoxide dismutase. Finally, peroxide is used by myeloperoxidase to oxidize chloride ions into hypochlorite (the active component of bleach), which is toxic to bacteria. Thus, NADPH oxidase is critical for phagocyte killing of bacteria through reactive oxygen species.

(Two other mechanisms are used by phagocytes to kill bacteria: nitric oxide and proteases, but the loss of ROS-mediated killing alone is sufficient to cause chronic granulomatous disease.)

Defects in one of the four essential subunits of phagocyte NADPH oxidase (PHOX) can all cause CGD of varying severity, dependent on the defect. There are over 410 known possible defects in the PHOX enzyme complex that can lead to chronic granulomatous disease.

Diagnosis

The nitroblue-tetrazolium (NBT) test is the original and most widely known test for chronic granulomatous disease. It is negative in CGD, meaning that it does not turn blue. The higher the blue score, the better the cell is at producing reactive oxygen species. This test depends upon the direct reduction of NBT to the insoluble blue compound formazan by NADPH oxidase; NADPH is oxidized in the same reaction. This test is simple to perform and gives rapid results, but only tells whether or not there is a problem with the PHOX enzymes, not how much they are affected. A similar test uses dihydrorhodamine (DHR) where whole blood is stained with DHR, incubated, and stimulated to produce superoxide radicals which oxidize DHR to rhodamin in cells with normal function. An advanced test called the cytochrome C reduction assay tells physicians how much superoxide a patient's phagocytes can produce. Once the diagnosis of CGD is established, a genetic analysis may be used to determine exactly which mutation is the underlying cause.

Treatment

Management of chronic granulomatous disease revolves around two goals: 1) diagnose the disease early so that antibiotic prophylaxis can be given to keep an infection from occurring, and 2) educate the patient about his or her condition so that prompt treatment can be given if an infection occurs.

Antibiotics

Physicians often prescribe the antibiotic trimethoprim-sulfamethoxazole to prevent bacterial infections. This drug also has the benefit of sparing the normal bacteria of the digestive tract. Fungal infection is commonly prevented with itraconazole, although a newer drug of the same type called voriconazole may be more effective. The use of this drug for this purpose is still under scientific investigation.

Immunomodulation

Interferon, in the form of interferon gamma-1b (Actimmune) is approved by the Food and Drug Administration for the prevention of infection in CGD. It has been shown to reduce infections in CGD patients by 70% and to decrease their severity. Although its exact mechanism is still not entirely understood, it has the ability to give CGD patients more immune function and therefore, greater ability to fight off infections. This therapy has been standard treatment for CGD for several years.

Hematopoietic Stem Cell Transplantation (HSCT)

Hematopoietic stem cell transplantation from a matched donor is curative although not without significant risk.

Prognosis

There are currently no studies detailing the long term outcome of chronic granulomatous disease with modern treatment. Without treatment, children often die in the first decade of life. The increased severity of X-linked CGD results in a decreased survival rate of patients, as 20% of X-linked patients die of CGD-related causes by the age of 10, whereas 20% of autosomal recessive patients die by the age of 35. Recent experience from centers specializing in the care of patients with CGD suggests that the current mortality has fallen to under 3% and 1% respectively. CGD was initially termed "fatal granulomatous disease of childhood" because patients rarely survived past their first decade in the time before routine use of prophylactic antimicrobial agents. The average patient now survives at least 40 years.

Epidemiology

CGD affects about 1 in 200,000 people in the United States, with about 20 new cases diagnosed each year.

Chronic granulomatous disease affects all people of all races, however, there is limited information on prevalence outside of the United States. One survey in Sweden reported an incidence of 1 in 220,000 people, while a larger review of studies in Europe suggested a lower rate: 1 in 250,000 people.

History

This condition was first described in 1954 by Janeway, who reported five cases of the disease in children. In 1957 it was further characterized as "a fatal granulomatosus of childhood". The underlying cellular mechanism that causes chronic granulomatous disease was discovered in 1967, and research since that time has further elucidated the molecular mechanisms underlying the disease. Use of antibiotic prophylaxis, surgical abscess drainage, and vaccination led to the term "fatal" being dropped from the name of the disease as children survived into adulthood.

Research

Gene therapy is currently being studied as a possible treatment for chronic granulomatous disease. CGD is well-suited for gene therapy since it is caused by a mutation in single gene which only affects one body system (the hematopoietic system). Viruses have been used to deliver a normal gp91 gene to rats with a mutation in this gene, and subsequently the phagocytes in these rats were able to produce oxygen radicals.

In 2006, two human patients with X-linked chronic granulomatous disease underwent gene therapy and blood cell precursor stem cell transplantation to their bone marrow. Both patients recovered from their CGD, clearing pre-existing infections and demonstrating increased oxidase activity in their neutrophils. However, long-term complications and efficacy of this therapy were unknown.

In 2012, a 16-year-old boy with CGD was treated at the Great Ormond Street Hospital, London with an experimental gene therapy which temporarily reversed the CGD and allowed him to overcome a life-threatening lung disease.

References

- Harrison's Principles of Internal Medicine: Volumes 1 and 2, 18th Edition (18 ed.). McGraw-Hill Professional. 2011-08-11. ISBN 9780071748896.

- MacKay, edited by Noel Richard Rose, Ian R. (2014). The autoimmune diseases (Fifth ed.). [S.l.]: Academic Press. p. Chapter 75. ISBN 978-0-12-384929-8.

- Kumar, Vinay; Abbas, Abul K.; Aster, Jon C., eds. (2014). "Hypersensitivity: Immunologicaly Mediated Tissue Injury". Robbins & Cotran Pathologic Basis of Disease (9th ed.). Elsevier Health Sciences. pp. 200–11. ISBN 978-0-323-29635-9.

- Mitchell, Richard Sheppard; Kumar, Vinay; Abbas, Abul K.; Fausto, Nelson (2007). "Table 5-1". Robbins Basic Pathology (8th ed.). Philadelphia: Saunders. ISBN 1-4160-2973-7.

- Markowitz, edited by William N. Rom ; associate editor, Steven B. (2007). Environmental and occupational medicine (4th ed.). Philadelphia: Wolters Kluwer/Lippincott Williams & Wilkins. p. 745. ISBN 978-0-7817-6299-1.

- Guideline on when to start antiretroviral therapy and on pre-exposure prophylaxis for HIV. (PDF). WHO. 2015. p. 13. ISBN 9789241509565.

- "The impact of AIDS on people and societies" (PDF). 2006 Report on the global AIDS epidemic. UNAIDS. 2006. ISBN 92-9173-479-9. Retrieved June 14, 2006.

- Evian, Clive (2006). Primary HIV/AIDS care: a practical guide for primary health care personnel in a clinical and supportive setting (Updated 4th ed.). Houghton [South Africa]: Jacana. p. 29. ISBN 978-1-77009-198-6.

- Charles B. Hicks, MD (2001). Jacques W. A. J. Reeders & Philip Charles Goodman, ed. Radiology of AIDS. Berlin [u.a.]: Springer. p. 19. ISBN 978-3-540-66510-6.

- Cunha, Burke (2012). Antibiotic Essentials 2012 (11 ed.). Jones & Bartlett Publishers. p. 303. ISBN 9781449693831.

- Stürchler, Dieter A. (2006). Exposure a guide to sources of infections. Washington, DC: ASM Press. p. 544. ISBN 978-1-55581-376-5.

- Kerrigan, Deanna (2012). The Global HIV Epidemics among Sex Workers. World Bank Publications. p. 1. ISBN 978-0-8213-9775-6.

- Aral, Sevgi (2013). The New Public Health and STD/HIV Prevention: Personal, Public and Health Systems Approaches. Springer. p. 120. ISBN 978-1-4614-4526-5.

- Martínez, edited by Miguel Angel (2010). RNA interference and viruses : current innovations and future trends. Norfolk: Caister Academic Press. p. 73. ISBN 978-1-904455-56-1.

- Pillay, Deenan; Genetti, Anna Maria; Weiss, Robin A. (2007). "Human Immunodeficiency Viruses". In Zuckerman, Arie J.; et al. Principles and practice of clinical virology (6th ed.). Hoboken, N.J.: Wiley. p. 905. ISBN 978-0-470-51799-4.

6

Diagnosis and Treatment Related to Immunological Disorders

There are a plethora of options to treat immunological disorders. This chapter discusses the topic of immunodiagnostics that utilizes immunoassays to measure the presence or concentration of a macromolecule or a small molecule in a solution through the use of an antigen or an antibody. Another method of diagnostics is immunotherapy that is either designed to elicit /amplify the immune response or to reduce/ suppress an immune response as the situation calls for. The aspects of each option elucidated in the chapter are of vital importance and promote a better understanding of the topic of the diagnosis and treatment of immunological disorders.

Immunodiagnostics

Immunodiagnostics is a diagnostic methodology that uses an antigen-antibody reaction as their primary means of detection. The concept of using immunology as a diagnostic tool was introduced in 1960 as a test for serum insulin. A second test was developed in 1970 as a test for thyroxine in the 1970s.

It is well-suited for the detection of even the smallest of amounts of (bio)chemical substances. Antibodies specific for a desired antigen can be conjugated with a radiolabel, fluorescent label, or color-forming enzyme and are used as a "probe" to detect it. Well known applications include pregnancy tests, immunoblotting, ELISA and immunohistochemical staining of microscope slides. The speed, accuracy and simplicity of such tests has led to the development of rapid techniques for the diagnosis of disease, microbes and even illegal drugs *in vivo* (of course tests conducted in a closed environment have a higher degree of accuracy). Such testing is also used to distinguish compatible blood types.

The Enzyme-Linked ImmunoSorbent Assay or ELISA and the Lateral-Flow test, also known as the dipstick or rapid test, currently are the two predominant formats in immunodiagnostics.

The ELISA

The ELISA (sometimes also called an EIA) is a sensitive, inexpensive assay technique involving the use of antibodies coupled with indicators (e.g. enzymes linked to dyes) to detect the presence of specific substances, such as enzymes, viruses, or bacteria. While there are several different types, basically ELISAs are created by coating a suitable plastic (the solid phase) with an antibody. To complete the reaction, a sample believed to contain the antigen of interest is added to the solid phase. Then a second antibody coupled with an enzyme is used followed by the addition of a color-forming substrate specific to the antibody.

Immunotherapy

Immunotherapy is the "treatment of disease by inducing, enhancing, or suppressing an immune response". Immunotherapies designed to elicit or amplify an immune response are classified as activation immunotherapies, while immunotherapies that reduce or suppress are classified as suppression immunotherapies.

Immunomodulatory regimens often have fewer side effects than existing drugs, including less potential for creating resistance in microbial disease.

Cell-based immunotherapies are effective for some cancers. Immune effector cells such as lymphocytes, macrophages, dendritic cells, natural killer cells (NK Cell), cytotoxic T lymphocytes (CTL), etc., work together to defend the body against cancer by targeting abnormal antigens expressed on the surface of tumor cells.

Therapies such as granulocyte colony-stimulating factor (G-CSF), interferons, imiquimod and cellular membrane fractions from bacteria are licensed for medical use. Others including IL-2, IL-7, IL-12, various chemokines, synthetic cytosine phosphate-guanosine (CpG) oligodeoxynucleotides and glucans are involved in clinical and preclinical studies.

Immunomodulators

Immunomodulators are the active agents of immunotherapy. They are a diverse array of recombinant, synthetic and natural preparations.

Class	Example agents
Interleukins	IL-2, IL-7, IL-12
Cytokines	Interferons, G-CSF, Imiquimod
Chemokines	CCL3, CCL26, CXCL7
Immunomodulatory imide drugs (IMiDs)	thalidomide and its analogues (lenalidomide, pomalidomide, and apremilast)
Other	cytosine phosphate-guanosine, oligodeoxynucleotides, glucans

Activation Immunotherapies

Cancer

Cancer immunotherapy attempts to stimulate the immune system to destroy tumors. A variety of strategies are in use or are undergoing research and testing. Randomized controlled studies in different cancers resulting in significant increase in survival and disease free period have been reported and its efficacy is enhanced by 20–30% when cell-based immunotherapy is combined with conventional treatment methods.

The extraction of G-CSF lymphocytes from the blood and expanding in vitro against a tumour antigen before reinjecting the cells with appropriate stimulatory cytokines. The cells then destroy the tumor cells that express the antigen.

BCG immunotherapy for early stage (non-invasive) bladder cancer instills attenuated live bacteria into the bladder and is effective in preventing recurrence in up to two thirds of cases.

Topical immunotherapy utilizes an immune enhancement cream (imiquimod) which produces interferon, causing the recipient's killer T cells to destroy warts, actinic keratoses, basal cell cancer, vaginal intraepithelial neoplasia, squamous cell cancer, cutaneous lymphoma, and superficial malignant melanoma.

Injection immunotherapy ("intralesional" or "intratumoral") uses mumps, candida, the HPV vaccine or trichophytin antigen injections to treat warts (HPV induced tumors).

Adoptive cell transfer has been tested on lung and other cancers.

Dendritic Cell-Based Pump-Priming

Dendritic cells can be stimulated to activate a cytotoxic response towards an antigen. Dendritic cells, a type of antigen presenting cell, are harvested from the person needing the immunotherapy. These cells are then either pulsed with an antigen or tumor lysate or transfected with a viral vector, causing them to display the antigen. Upon transfusion into the person, these activated cells present the antigen to the effector lymphocytes (CD4+ helper T cells, cytotoxic CD8+ T cells and B cells). This initiates a cytotoxic response against tumor cells expressing the antigen (against which the adaptive response has now been primed). The cancer vaccine Sipuleucel-T is one example of this approach.

T-Cell Adoptive Transfer

Adoptive cell transfer *in vitro* cultivates autologous, extracted T cells for later transfusion. The T cells may already target tumor cells. Alternatively, they may be genetically engineered to do so. These T cells, referred to as tumor-infiltrating lymphocytes (TIL), are multiplied using high concentrations of Interleukin-2, anti-CD3 and allo-reactive feeder cells. These T cells are then transferred back into the person along with administration of IL-2 to further boost their anti-cancer activity.

Before reinfusion, lymphodepletion of the recipient is required to eliminate regulatory T cells as well as unmodified, endogenous lymphocytes that compete with the transferred cells for homeostatic cytokines. Lymphodepletion can be achieved by total body irradiation. Transferred cells multiplied *in vivo* and persisted in peripheral blood in many people, sometimes representing levels of 75% of all CD8+ T cells at 6–12 months after infusion. As of 2012, clinical trials for metastatic melanoma were ongoing at multiple sites.

Immune Enhancement Therapy

Autologous immune enhancement therapy use the person's own peripheral blood-derived natural killer cells, cytotoxic T lymphocytes and other relevant immune cells are expanded *in vitro* and then reinfused. The therapy has been tested against Hepatitis C, Chronic fatigue syndrome and HHV6 infection.

Genetically Engineered T Cells

Genetically engineered T cells are created by harvesting T cells and then infecting the T cells with a retrovirus that contains a copy of a T cell receptor (TCR) gene that is specialised to recognise tumour antigens. The virus integrates the receptor into the T cells' genome. The cells are expanded non-specifically and/or stimulated. The cells are then reinfused and produce an immune response against the tumour cells. The technique has been tested on refractory stage IV metastatic melanomas and advanced skin cancer

Immune Recovery

Another potential use of immunotherapy is to restore the immune system of people with immune deficiencies. Cytokines, Interleukin-7 and Interleukin-2 have been tested in clinical trials.

Vaccination

Antimicrobial immunotherapy, which includes vaccination, involves activating the immune system to respond to an infectious agent.

Suppression Immunotherapies

Immune suppression dampens an abnormal immune response in autoimmune diseases or reduces a normal immune response to prevent rejection of transplanted organs or cells.

Immunosuppressive Drugs

Immunosuppressive drugs help manage organ transplantation and autoimmune disease. Immune responses depend on lymphocyte proliferation. Cytostatic drugs are immunosuppressive. Glucocorticoids are somewhat more specific inhibitors of lymphocyte activation, whereas inhibitors of immunophilins more specifically target T lymphocyte activation. Immunosuppressive antibodies target steps in the immune response. Other drugs modulate immune responses.

Immune Tolerance

The body naturally does not launch an immune system attack on its own tissues. Immune tolerance therapies seek to reset the immune system so that the body stops mistakenly attacking its own organs or cells in autoimmune disease or accepts foreign tissue in organ transplantation. Creating immunity reduces or eliminates the need for lifelong immunosuppression and attendant side effects. It has been tested on transplantations, and type 1 diabetes or other autoimmune disorders.

Allergies

Immunotherapy is used to treat allergies. While allergy treatments (such as antihistamines or corticosteroids) treat allergic symptoms, immunotherapy can reduce sensitivity to allergens, lessening its severity.

Immunotherapy may produce long-term benefits. Immunotherapy is partly effective in some people and ineffective in others, but it offers allergy sufferers a chance to reduce or stop their symptoms.

The therapy is indicated for people who are extremely allergic or who cannot avoid specific allergens. Immunotherapy is generally not indicated for food or medicinal allergies. This therapy is particularly useful for people with allergic rhinitis or asthma.

The first dose contain tiny amounts of the allergen or antigen. Dosages increase over time, as the person becomes desensitized. This technique has been tested on infants to prevent peanut allergies.

Helminthic Therapies

Whipworm ova (*Trichuris suis*) and Hookworm (*Necator americanus*) have been tested for immunological diseases and allergies. Helminthic therapy has been investigated as a treatment for relapsing remitting multiple sclerosis Crohn's, allergies and asthma. The mechanism of how the helminths modulate the immune response, is unknown. Hypothesized mechanisms include re-polarisation of the Th1 / Th2 response and modulation of dendritic cell function. The helminths down regulate the pro-inflammatory Th1 cytokines, Interleukin-12 (IL-12), Interferon-Gamma (IFN-γ) and Tumour Necrosis Factor-Alpha (TNF-ἀ), while promoting the production of regulatory Th2 cytokines such as IL-10, IL-4, IL-5 and IL-13.

Co-evolution with helminths has shaped some of the genes associated with Interleukin expression and immunological disorders, such Crohn's, ulcerative colitis and Celiac disease. Helminth's relationship to humans as hosts should be classified as mutualistic or symbiotic.

Vaccination

Vaccination is the administration of antigenic material (a vaccine) to stimulate an individual's immune system to develop adaptive immunity to a pathogen. Vaccines can prevent or ameliorate morbidity from infection. When a sufficiently large percentage of a population has been vaccinated, this results in herd immunity. The effectiveness of vaccination has been widely studied and verified; for example, the influenza vaccine, the HPV vaccine, and the chicken pox vaccine. Vaccination is the most effective method of preventing infectious diseases; widespread immunity due to vaccination is largely responsible for the worldwide eradication of smallpox and the restriction of diseases such as polio, measles, and tetanus from much of the world. The World Health Organization (WHO) reports that licensed vaccines are currently available to prevent or contribute to the prevention and control of twenty-five preventable infections.

The active agent of a vaccine may be intact but inactivated (non-infective) or attenuated (with reduced infectivity) forms of the causative pathogens, or purified components of the pathogen that have been found to be highly immunogenic (e.g., outer coat proteins of a virus). Toxoids are produced for immunization against toxin-based diseases, such as the modification of tetanospasmin toxin of tetanus to remove its toxic effect but retain its immunogenic effect.

Smallpox was most likely the first disease people tried to prevent by inoculating themselves and was the first disease for which a vaccine was produced. The smallpox vaccine was discovered in 1796 by the British physician Edward Jenner, although at least six people had used the same principles years earlier. Louis Pasteur furthered the concept through his work in microbiology. The immunization was called *vaccination* because it was derived from a virus affecting cows (Latin: *vacca—cow*). Smallpox was a contagious and deadly disease, causing the deaths of 20–60% of infected adults and over 80% of infected children. When smallpox was finally eradicated in 1979, it had already killed an estimated 300–500 million people in the 20th century.

In common speech, *vaccination* and *immunization* have a similar meaning. This distinguishes it from inoculation, which uses unweakened live pathogens, although in common usage either can refer to an immunization. Vaccination efforts have been met with some controversy on scientific, ethical, political, medical safety, and religious grounds. In rare cases, vaccinations can injure people and, in the United States, they may receive compensation for those injuries under the National Vaccine Injury Compensation Program. Early success and compulsion brought widespread acceptance, and mass vaccination campaigns have greatly reduced the incidence of many diseases in numerous geographic regions.

Mechanism of Function

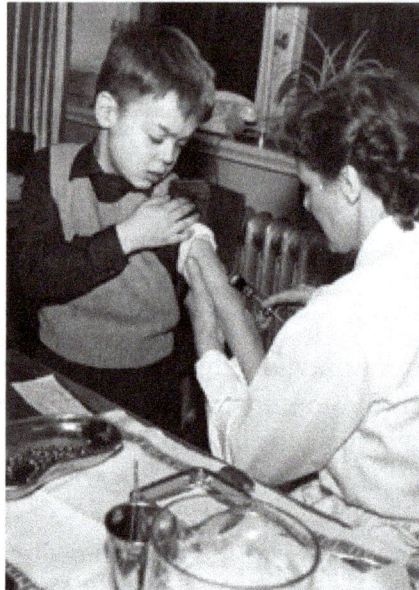

Polio vaccination started in Sweden in 1957.

Generically, the process of artificial induction of immunity, in an effort to protect against infectious disease, works by 'priming' the immune system with an 'immunogen'. Stimulating immune responses with an infectious agent is known as *immunization*. Vaccination includes various ways of administering immunogens.

Some vaccines are administered after the patient already has contracted a disease. Vaccines given after exposure to smallpox, within the first three days, are reported to attenuate the disease considerably, and vaccination up to a week after exposure probably offers some protection from disease or may modify the severity of disease. The first rabies immunization was given by Louis Pasteur to a child after he was bitten by a rabid dog. Subsequent to this, it has been found that, in people with uncompromised immune systems, four doses of rabies vaccine over 14 days, wound care, and treatment of the bite with rabies immune globulin, commenced as soon as possible after exposure, is effective in preventing the development of rabies in humans. Other examples include experimental AIDS, cancer and Alzheimer's disease vaccines. Such immunizations aim to trigger an immune response more rapidly and with less harm than natural infection.

Most vaccines are given by hypodermic injection as they are not absorbed reliably through the intestines. Live attenuated polio, some typhoid, and some cholera vaccines are given orally to produce immunity in the bowel. While vaccination provides a lasting effect, it usually takes several weeks to develop, while passive immunity (the transfer of antibodies) has immediate effect.

Adjuvants and Preservatives

Vaccines typically contain one or more adjuvants, used to boost the immune response. Tetanus toxoid, for instance, is usually adsorbed onto alum. This presents the antigen in such a way as to produce a greater action than the simple aqueous tetanus toxoid. People who get an excessive reaction to adsorbed tetanus toxoid may be given the simple vaccine when time for a booster occurs.

In the preparation for the 1990 Persian Gulf campaign, pertussis vaccine (not acellular) was used as an adjuvant for anthrax vaccine. This produces a more rapid immune response than giving only the anthrax, which is of some benefit if exposure might be imminent.

Vaccines may also contain preservatives to prevent contamination with bacteria or fungi. Until recent years, the preservative thimerosal was used in many vaccines that did not contain live virus. As of 2005, the only childhood vaccine in the U.S. that contains thimerosal in greater than trace amounts is the influenza vaccine, which is currently recommended only for children with certain risk factors. Single-dose influenza vaccines supplied in the UK do not list thiomersal (its UK name) in the ingredients. Preservatives may be used at various stages of production of vaccines, and the most sophisticated methods of measurement might detect traces of them in the finished product, as they may in the environment and population as a whole.

Vaccination Versus Inoculation

Many times these words are used interchangeably, as if they were synonyms. In fact, they are different things. As doctor Byron Plant explains: "Vaccination is the more commonly used term, which actually consists of a 'safe' injection of a sample taken from a cow suffering from cowpox... Inoculation, a practice probably as old as the disease itself, is the injection of the variola virus taken from a pustule or scab of a smallpox sufferer into the superficial layers of the skin, commonly on the upper arm of the subject. Often inoculation was done 'arm to arm' or less effectively 'scab to arm'..."

Vaccination began in the 18th century with the work of Edward Jenner.

Types

Vaccines work by presenting a foreign antigen to the immune system to evoke an immune response, but there are several ways to do this. Four main types are currently in clinical use:

1. An inactivated vaccine consists of virus or bacteria that are grown in culture and then killed using a method such as heat or formaldehyde. Although the virus or bacteria particles are destroyed and cannot replicate, the virus capsid proteins or bacterial wall are intact enough

to be recognized and remembered by the immune system and evoke a response. When manufactured correctly, the vaccine is not infectious, but improper inactivation can result in intact and infectious particles. Since the properly produced vaccine does not reproduce, booster shots are required periodically to reinforce the immune response.

2. In an attenuated vaccine, live virus or bacteria with very low virulence are administered. They will replicate, but locally or very slowly. Since they do reproduce and continue to present antigen to the immune system beyond the initial vaccination, boosters may be required less often. These vaccines may be produced by passaging, for example, adapting a virus into different host cell cultures, such as in animals, or at suboptimal temperatures, allowing selection of less virulent strains, or by mutagenesis or targeted deletions in genes required for virulence. There is a small risk of reversion to virulence, which is smaller in vaccines with deletions. Attenuated vaccines also cannot be used by immunocompromised individuals. Reversions of virulence were described for a few attenuated viruses of chickens (infectious bursal disease virus, avian infectious bronchitis virus, avian infectious laryngotracheitis virus , avian metapneumovirus)

3. Virus-like particle vaccines consist of viral protein(s) derived from the structural proteins of a virus. These proteins can self-assemble into particles that resemble the virus from which they were derived but lack viral nucleic acid, meaning that they are not infectious. Because of their highly repetitive, multivalent structure, virus-like particles are typically more immunogenic than subunit vaccines (described below). The human papillomavirus and Hepatitis B virus vaccines are two virus-like particle-based vaccines currently in clinical use.

4. A subunit vaccine presents an antigen to the immune system without introducing viral particles, whole or otherwise. One method of production involves isolation of a specific protein from a virus or bacterium (such as a bacterial toxin) and administering this by itself. A weakness of this technique is that isolated proteins may have a different three-dimensional structure than the protein in its normal context, and will induce antibodies that may not recognize the infectious organism. In addition, subunit vaccines often elicit weaker antibody responses than the other classes of vaccines.

A number of other vaccine strategies are under experimental investigation. These include DNA vaccination and recombinant viral vectors.

History

It is known that the process of inoculation was used by Chinese physicians in the 10th century. Scholar Ole Lund comments: "The earliest documented examples of vaccination are from India and China in the 17th century, where vaccination with powdered scabs from people infected with smallpox was used to protect against the disease. Smallpox used to be a common disease throughout the world and 20 to 30% of infected persons died from the disease. Smallpox was responsible for 8 to 20% of all deaths in several European countries in the 18th century. The tradition of vaccination may have originated in India in AD 1000." The mention of inoculation in the *Sact'eya Grantham*, an Ayurvedic text, was noted by the French scholar Henri Marie Husson in the journal *Dictionaire des sciences médicales*. Inoculation was reportedly widely practised in China in the reign of the Longqing Emperor (r. 1567–1572) during the Ming Dynasty (1368–1644). The Anatolian Ottoman Turks knew about methods of inoculation. This kind of inoculation and other forms of variolation were introduced into England by Lady Montagu, a famous English letter-writer and wife of the English ambassador at Istanbul between 1716 and 1718, who almost died from smallpox as a young adult and was physically scarred from it. Inoculation was adopted both in England and in America nearly half a century before Jenner's famous smallpox vaccine of 1796 but the death rate of about 2% from this method meant that it was mainly used during dangerous outbreaks of the disease and remained controversial.

Jenner's handwritten draft of the first vaccination

It was noticed during the 18th century that people who had suffered from the less virulent cowpox were immune to smallpox and the first recorded use of this idea was by a farmer Benjamin Jesty at Yetminster who had suffered the disease and transmitted it to his own family in 1774, his sons subsequently not getting the mild version of smallpox when later inoculated in 1789. But it was Edward Jenner, a doctor in Berkeley, who established the procedure by introducing material from a cowpox vesicle on Sarah Nelmes, a milkmaid, into the arm of a boy named James Phipps. Two months later he inoculated the boy with smallpox and the disease did not develop. In 1798, Jenner published "An Inquiry into the Causes and Effects of the Variolae Vacciniae" which coined the term *vaccination* and created widespread interest. He distinguished 'true' and 'spurious' cowpox (which did not give the desired effect) and developed an "arm-to-arm" method of propagating the vaccine from the vaccinated individual's pustule. Early attempts at confirmation were confounded by contamination with smallpox, but despite controversy within the medical profession and religious opposition to the use of animal material, by 1801 his report was translated into six languages and over 100,000 people were vaccinated.

Since then vaccination campaigns have spread throughout the globe, sometimes prescribed by law or regulations. Vaccines are now used against a wide variety of diseases besides smallpox. Louis Pasteur further developed the technique during the 19th century, extending its use to killed agents protecting against anthrax and rabies. The method Pasteur used entailed treating the agents for those diseases so they lost the ability to infect, whereas inoculation was the hopeful selection of a less virulent form of the disease, and Jenner's vaccination entailed the substitution of a different and less dangerous disease for the one protected against. Pasteur adopted the name *vaccine* as a generic term in honor of Jenner's discovery.

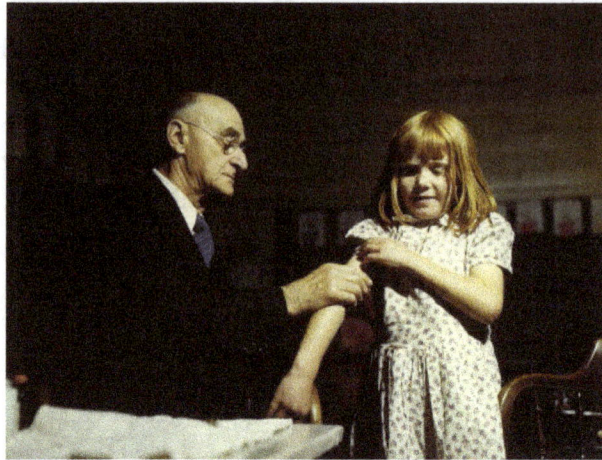

A doctor performing a typhoid vaccination in Texas, 1943

Maurice Hilleman was the most prolific vaccine inventor, and developed successful vaccines for measles, mumps, hepatitis A, hepatitis B, chickenpox, meningitis, pneumonia and *Haemophilus influenzae*.

In modern times, the first vaccine-preventable disease targeted for eradication was smallpox. The World Health Organization (WHO) coordinated this global eradication effort. The last naturally occurring case of smallpox occurred in Somalia in 1977. In 1988, the governing body of WHO targeted polio for eradication by 2000. Although the target was missed, eradication is very close. The next disease to be targeted for eradication would most likely be measles, which has declined since the introduction of measles vaccination in 1963.

In 2000, the Global Alliance for Vaccines and Immunization was established to strengthen routine vaccinations and introduce new and under-used vaccines in countries with a per capita GDP of under US$1000. GAVI is now entering its second phase of funding, which extends through 2014.

Society and Culture

To eliminate the risk of outbreaks of some diseases, at various times several governments and other institutions have employed policies requiring vaccination for all people. For example, an 1853 law required universal vaccination against smallpox in England and Wales, with fines levied on people who did not comply. Common contemporary U.S. vaccination policies require that children receive common vaccinations before entering public school.

Beginning with early vaccination in the nineteenth century, these policies were resisted by a variety of groups, collectively called antivaccinationists, who object on scientific, ethical, political, medical safety, religious, and other grounds. Common objections are that vaccinations do not work, that compulsory vaccination represents excessive government intervention in personal matters, or that the proposed vaccinations are not sufficiently safe. Many modern vaccination policies allow exemptions for people who have compromised immune systems, allergies to the components used in vaccinations or strongly held objections.

In countries with limited financial resources, limited coverage causes much morbidity and mortality. More affluent countries are able to subsidize vaccinations for at-risk groups, resulting in more comprehensive and effective coverage. In Australia, for example, the Government subsidizes vaccinations for seniors and indigenous Australians.

Public Health Law Research, an independent US based organization, reported in 2009 that there is insufficient evidence to assess the effectiveness of requiring vaccinations as a condition for specified jobs as a means of reducing incidence of specific diseases among particularly vulnerable populations; that there is sufficient evidence supporting the effectiveness of requiring vaccinations as a condition for attending child care facilities and schools; and that there is strong evidence supporting the effectiveness of standing orders, which allow healthcare workers without prescription authority to administer vaccine as a public health intervention aimed at increasing vaccination rates.

Vaccination-Autism Controversy

In the MMR vaccine controversy, a fraudulent 1998 paper by Andrew Wakefield, originally published in *The Lancet*, presented supposed evidence that the MMR vaccine (an immunization against measles, mumps and rubella that is typically first administered to children shortly after their first birthday) was linked to the onset of autism spectrum disorders. The article was widely criticized for lack of scientific rigour, partially retracted in 2004 by Wakefield's co-authors, and was fully retracted by *The Lancet* in 2010. Wakefield was struck off the UK's medical registry for the fraud.

This Lancet article has sparked a much greater anti-vaccination movement, primarily in the United States. Even though the article was fraudulent and was retracted, 1 in 4 parents still believe vaccines can cause autism. Many parents do not vaccinate their children because they feel that diseases are no longer present due to all the vaccinations. This is a false assumption, since some diseases could still return. These pathogens could possibly infect vaccinated people, due to the pathogen's ability to mutate when it is able to live in unvaccinated hosts. In 2010, there was a whooping cough outbreak in California that was the worst outbreak in 50 years. A possible contributing factor was parents choosing to exempt their children from vaccinations. There was also a case in Texas in 2012 where 21 members of a church contracted measles because they chose to abstain from immunizations.

Side Effects and Injury

The Centers for Disease Control and Prevention (CDC) has compiled a list of vaccines and their possible side effects. Allegations of vaccine injuries in recent decades have appeared in litigation in the U.S. Some families have won substantial awards from sympathetic juries, even though most public health officials have said that the claims of injuries were unfounded. In response, several vaccine makers stopped production, which the US government believed could be a threat to public health, so laws were passed to shield makers from liabilities stemming from vaccine injury claims.

Routes of Administration

A vaccine administration may be oral, by injection (intramuscular, intradermal, subcutaneous), by puncture, transdermal or intranasal. Several recent clinical trials have aimed to deliver the vaccines via mucosal surfaces to be up-taken by the common mucosal immunity system, thus avoiding the need for injections.

Air France Vaccinations Centre in the 7th arrondissement of Paris

Global Trends in Vaccination

The World Health Organization (WHO) estimate that vaccination averts 2-3 million deaths per year (in all age groups), and up to 1.5 million children die each year due to diseases which could have been prevented by vaccination. They estimate that 29% of deaths of children under five years old in 2013 were vaccine preventable.

• Vaccination in art

La vaccine or *Le préjugé vaincu* by Louis-Léopold Boilly, 1807

A doctor vaccinating a small girl, other girls with loosened blouses wait their turn apprehensively by Lance Calkin

References

- Brooker S, Bethony J, Hotez PJ (2004). "Human Hookworm Infection in the 21st Century". Advances in Parasitology. 58: 197–288. doi:10.1016/S0065-308X(04)58004-1. ISBN 9780120317585. PMC 2268732. PMID 15603764.

- Koplow, David A. (2003). Smallpox: the fight to eradicate a global scourge. Berkeley: University of California Press. ISBN 0-520-24220-3.

- Offit PA (2007). Vaccinated: One Man's Quest to Defeat the World's Deadliest Diseases. Washington, DC: Smithsonian. ISBN 0-06-122796-X.

- Plotkin, Stanley A. (2006). Mass Vaccination: Global Aspects - Progress and Obstacles (Current Topics in Microbiology & Immunology). Springer-Verlag Berlin and Heidelberg GmbH & Co. K. ISBN 978-3-540-29382-8.

- Lund, Ole; Nielsen, Morten Strunge and Lundegaard, Claus (2005). Immunological Bioinformatics. MIT Press. ISBN 0-262-12280-4

- ""Laws and Policies Requiring Specified Vaccinations among High Risk Populations". Public Health Law Research. 7 December 2009. Retrieved 2014-11-19.

- "Vaccination Requirements for Child Care, School and College Attendance". Public Health Law Research. 12 July 2009. Retrieved 2014-11-19.

- Di Lorenzo G, Buonerba C, Kantoff PW (September 2011). "Immunotherapy for the treatment of prostate cancer". Nature Reviews Clinical Oncology. 8 (9): 551–61. doi:10.1038/nrclinonc.2011.72. PMID 21606971.

Permissions

All chapters in this book are published with permission under the Creative Commons Attribution Share Alike License or equivalent. Every chapter published in this book has been scrutinized by our experts. Their significance has been extensively debated. The topics covered herein carry significant information for a comprehensive understanding. They may even be implemented as practical applications or may be referred to as a beginning point for further studies.

We would like to thank the editorial team for lending their expertise to make the book truly unique. They have played a crucial role in the development of this book. Without their invaluable contributions this book wouldn't have been possible. They have made vital efforts to compile up to date information on the varied aspects of this subject to make this book a valuable addition to the collection of many professionals and students.

This book was conceptualized with the vision of imparting up-to-date and integrated information in this field. To ensure the same, a matchless editorial board was set up. Every individual on the board went through rigorous rounds of assessment to prove their worth. After which they invested a large part of their time researching and compiling the most relevant data for our readers.

The editorial board has been involved in producing this book since its inception. They have spent rigorous hours researching and exploring the diverse topics which have resulted in the successful publishing of this book. They have passed on their knowledge of decades through this book. To expedite this challenging task, the publisher supported the team at every step. A small team of assistant editors was also appointed to further simplify the editing procedure and attain best results for the readers.

Apart from the editorial board, the designing team has also invested a significant amount of their time in understanding the subject and creating the most relevant covers. They scrutinized every image to scout for the most suitable representation of the subject and create an appropriate cover for the book.

The publishing team has been an ardent support to the editorial, designing and production team. Their endless efforts to recruit the best for this project, has resulted in the accomplishment of this book. They are a veteran in the field of academics and their pool of knowledge is as vast as their experience in printing. Their expertise and guidance has proved useful at every step. Their uncompromising quality standards have made this book an exceptional effort. Their encouragement from time to time has been an inspiration for everyone.

The publisher and the editorial board hope that this book will prove to be a valuable piece of knowledge for students, practitioners and scholars across the globe.

Index

www.ingramcontent.com/pod-product-compliance
Lightning Source LLC
Chambersburg PA
CBHW082025190326
41458CB00010B/3282